D0620812

The Spoils of Progress:
Environmental Pollution
in the Soviet Union

The MIT Press
Cambridge, Massachusetts, and
London, England

The Spoils of Progress:
Environmental Pollution
in the Soviet Union

Marshall I. Goldman

This book was designed by The MIT Press Design Department.
It was set in Monotype Baskerville
by Wolf Composition Co. Inc.
Printed by The Colonial Press Inc.
and bound by The Colonial Press Inc.
in the United States of America.

Library of Congress Cataloging in Publication Data

Goldman, Marshall I
  The spoils of progress

  Bibliography: p.
  1. Pollution—Russia. I. Title
TD187.5.R9G63   301.3′1′0947   71-39075
ISBN 0-262-07053-7

To Lake Baikal

Contents

Acknowledgments   ix

Introduction   1

1 Environmental Protection in the Socialist
  State—Theory and Law   9

2 The Economic and Political Propensity to
  Pollute—Reality   43

3 Water Pollution   77

4 Air Pollution   121

5 Abuses of Land and Raw Materials   151

6 The Pollution of Lake Baikal   177

7 Ecological Facelifting, or Improving Nature   211

8 Reshaping the Earth   239

9 Advantages of the Soviet System and
   International Implications of Soviet Policies   271

Appendix A Selected Laws on the Environment   293

Appendix B The Conservation Law of the Russian
   Republic, 1960   301

Appendix C Water Law, 1970   311

Appendix D Conversion Table   331

Bibliography   333

Index   359

Acknowledgments

A study like this, bridging several disciplines, would have been impossible without the help of many individuals and institutions spread out across the world. It is easy to know where to begin but hard to know where to stop in offering acknowledgments. My only concern is that I may omit someone or some group through absent-mindedness. If someone is slighted, it is out of inadvertence, not ingratitude.

To begin with, I am deeply grateful to Wellesley College for providing me with extremely generous financial support at a time of financial stringency. Financial help in one form or another for travel or for conferences during the course of the writing was also provided for or arranged by the Japanese government and Professor Shigeto Tsuru; the American Friends Service Committee, Laurama Pixton; the Finnish government, Hannu Halinen and Nicholas Polunin; the French government, International Social Science Council, Ecole Pratique des Hautes Etudes, and Professor Ignacy Sachs.

From the USSR, I have benefited particularly from discussions with Gregory Galazii, Gregory Khozin, Alexander Kaljadin, Gennadi Gerasimov, Joseph Livchak, Sergey Kop'ev, M. Loiter, Nikolai Chistiakov, and Vladimir Kunin, none of whom of course should be held responsible for any misinterpretation of their views.

Closer to home, I have had the advice of Leonard Kirsch, Leon Smolinski, Barney Schwalberg, Joseph Berliner, Abram Bergson, Susan Gardos, and Elena Vorobey of the Russian Research Center, James Fay of M.I.T., James Mackenzie of the Massachusetts Audubon Society, Gerald Parker of the Massachusetts Department of Public Health, Robert Campbell of the University of Indiana, Gregory Grossman of the University of California, and J. G. Tolpin of Northwestern University. Typing help has been provided by Rose di Benedetto, Ramsey Ives, Sarah Hoagland, and Alan Lotrick. The maps (except for Map 8.2) were drawn by Walter Minty, and the index was painstakingly prepared by Adele Wick.

Acknowledgment must also be made to the following journals for allowing me to reprint materials from the articles that originally appeared in their pages: "The Convergence of Environmental Disruption," in *Science*, vol. 170, October 2, 1970: "The Pollution of Lake Baikal," *The New Yorker*, June 19, 1971; and "Externali-

ties and the Race for Economic Growth in the USSR:
Will the Environment Ever Win?" the *Journal of
Political Economy*, March–April 1972.

Finally, as with everything else I write, the greatest
thanks go to my wife, Merle, who struggles and agonizes
over every page. As ungrateful as I may appear at the
time, I am boundlessly indebted to her for her patience,
persistence, and perception.

Introduction

By now it is a familiar story: rivers that blaze with fire, smog that suffocates cities, streams that vomit dead fish, oil slicks that blacken seacoasts, prized beaches that vanish in the waves, and lakes that evaporate and die a slow, smelly death. What makes it unfamiliar is that this is a description not only of the United States but also of the Soviet Union.

Most conservationists and social critics are unaware that the USSR has environmental disruption that is as extensive and severe as ours. Most of us have been so distressed by our own environmental disruption that we lack the emotional energy to worry about anyone else's difficulties. Yet it would further the understanding of our own situation and of possible solutions if we could explain why it is that a socialist or communist country like the USSR finds itself abusing the environment in the same way, and to the same degree, as we abuse it.

Not only are most of us unaware of environmental disruption in a socialist country like the USSR, but somehow it has become a tenet of conservationist folklore that environmental disruption will no longer exist in a society where the state owns all the means of production.[1] Most critics have automatically assumed that if all the factories in a society are state-owned, the state will ensure that the broader interests of the general public

[1] Wherever possible, the term environmental disruption is used in preference to pollution. We all seem to know what we mean by pollution, but often what is polluted to one species is perfectly suited to another. There is cause for concern however, when the environment is altered in such a way that the normal ecological processes are disrupted. For a fuller explanation see Chapter 5.

will be protected. Under the circumstances, presumably no factory will be allowed to emit untreated effluents. With the state as the owner, there would be no need to hold back on expenditures for pollution control since the public good, not vested interests or private profit, would be the sole determinant of action.

By contrast, pollution is frequently regarded as an inevitable by-product of capitalism. The tolerance, if not adulation, of private greed makes it all but certain that industrialists will pursue their own interests without regard for the public good. Since it is unlikely to increase their profits, most businessmen and their stockholders are thought to oppose expenditures for pollution control. If by chance a state-owned institution in the Soviet Union is caught polluting, Soviet authorities usually explain it away as a legacy of the capitalist system or a consequence of the destruction suffered in World War II. Supposedly, it will be only a matter of time before the inexorable logic and the institutions of public ownership induce the Soviet manager to act in a more selfless, less destructive manner.

Whether or not the reasoning behind such arguments is valid, the fact remains that private enterprise and the profit motive do undoubtedly account for a good portion of the environmental disruption that we encounter in the United States. Those who insist on the opposite, that environmental disruption is not the result of capitalism and private enterprise but of malfunctioning markets and government interference are even more hard put to prove their case (Allen 1970, p. 1). Most authorities agree that the vast majority of private businessmen have probably neglected the needs of the environment for the sake of their own immediate profit. The widespread acceptance of this fact by Marxists and non-Marxists, and radicals and conservatives in general, helps to explain the attention devoted to the effects of pollution in capitalist societies. Accordingly, for most critics, especially those ignorant of the actual situation in the Soviet Union, the solution is simple: increase the role of the government, preferably through regulation and even nationalization. Environmental disruption should then be reduced if not eliminated. For this reason, a study of environmental disruption in the USSR is a prerequisite for anyone who believes that it is capitalism and private greed that are the root cause of environmental disruption.

The first step in a study of environmental disruption in the Soviet Union is to understand their theoretical justification for public ownership as a means of eliminating pollution. Backing for such arguments comes not only from the socialist or communist theorists but from bourgeois theorists as well. Within the USSR the ideological framework has been solidly buttressed with comprehensive legislation that spells out the legal obligations of all industries and citizens toward preserving the environment. This will be the focus of Chapter 1.

Offsetting the ideology and formal legislation, however, is another set of forces, partly economic, partly political, and partly technical. These factors, as we shall discover in Chapter 2, lead the Russians in the opposite direction so that there are actually pressures to pollute in some instances and ineffective barriers in others. On balance, present institutional or legal arrangements in the USSR have not been able to stem environmental disruption. Of course, in a country as large as the USSR there are many places that have been spared man's disruptive incursions, but as the population grows in numbers and mobility, such areas become fewer and fewer. Moreover, as in the United States, the most idyllic sites are the very ones that tend to attract the Soviet population.

Just because human beings intrude on an area, it does not necessarily follow that the area's resources will be abused. Certainly the presence of human beings means some alteration in the previous ecological balance, and in some cases there may be severe damage, but the change need not always be a serious one. Nevertheless, many of the environmental changes that have taken place in the Soviet Union have been major in scope. As a result, the quality of the air, water, and land resources has been adversely affected.

Much of the pollution in the USSR has little to do with the prevailing form of ideology or government but much to do with the advent of industrialization and the consequences of rapid economic growth. Industrialization and urbanization have been at least as demanding on the environment in the USSR as they have been elsewhere in the developed world, and the result is virtually the same. The increasing output per capita, the discovery and manufacture of synthetic compounds, and the ever-increasing concentration of people in urban areas have ruptured traditional ecological systems in the USSR.

Instead of relieving themselves here and there over the vast expanse of the Soviet Union, the majority of Soviet citizens like more and more people all over the developed world are now flushing their toilets next to one another in large urban centers. No longer can human and industrial wastes be disposed of in "nature's way" by allowing them to break up and percolate through the soil, water, and atmosphere. Once human, industrial, and even agricultural wastes are concentrated, they overwhelm nature's normal ecological processes. Furthermore, a poor country engaged in a fight for industrialization finds it very difficult to divert money to electrostatic precipitators and away from more electric power stations. Environmental control is no easier when the economy begins to grow at a rapid rate. Be they adequate or inadequate, any prevention and treatment facilities that may have been provided before are soon swamped as production and waste discharge surge. This is as true in the Soviet Union as it is in Japan. Like people of other countries, the Russians now find themselves in the unenviable position where they need an increase in their rate of growth to make it possible for them to generate enough resources to provide for pollution control.

In addition, because investment and production decisions are determined by a small group of political officials and economic planners, economic growth in the USSR is often unbalanced. Occasionally, therefore, change takes place so rapidly and on such a massive scale in one area of the economy that there is no time to reflect on all the consequences. In the early 1960s, Khrushchev decided that the Soviet Union needed a large chemical industry. All at once chemical plants began to spring up all over the country. In their anxiety to fulfill their targets for new plant construction, few if any of the planners were able to devote much attention to the disruptive effects on the environment that such plants might have.

The effects of Soviet urbanization, industrialization, and growing technological prowess on the water, air, and land resources of the country will be examined in detail in Chapters 3 through 5. This discussion is followed in Chapters 6 through 8 by case studies of how Lake Baikal and the Caspian and Aral Seas have been exploited for economic purposes to the detriment of the environment. Finally, the genuine advantages of the Soviet system are considered in Chapter 9 along with some of the implications of Soviet practices for the world at large.

In sum, based on Soviet experience, there is no reason to believe that state ownership of the means of production will necessarily guarantee the elimination of environmental disruption. Industrialization, urbanization, and technological change give rise to environmental upheavals that so far no society has been able to harness completely. While state ownership of a country's productive resources may eliminate some forms of environmental disruption, it also exacerbates rather than ameliorates other forms. Were the United States and its form of government to vanish from the earth, the world would still have to contend with the Soviet supersonic transport and the degradation of Lake Baikal.

The relative merits for the environment of state ownership versus mixed or private ownership of the means of production are hard to compare since there seems to be no feasible way to decide whether American or Soviet pollution is more serious. One thing does seem clear, however: shortcut solutions like nationalization are hardly likely to eliminate pollution on our planet. Certain forms of government supervision and regulation are necessary to reduce environmental disruption, but that in itself is far from enough. Unfortunately there is no simple solution.

1 Environmental Protection in the Socialist State— Theory and Law

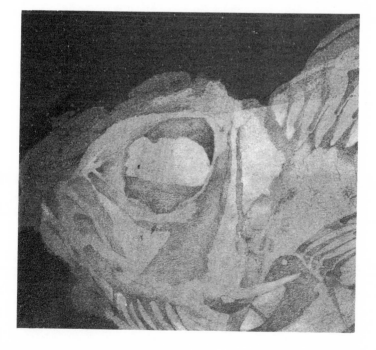

Conventional wisdom treats state ownership of the means of production as an effective guarantee against environmental disruption. After we trace the development of the ideology and theory of this conception, we can then analyze the type of environmental legislation adopted in the USSR. Having done this, we will have a background against which we can judge the effectiveness of environmental protection in a socialist or communist society.

Private Enterprise as the Chief Polluter
Why is it that when most people think of pollution, they instinctively tend to think of private enterprise as the prime offender? Much of this reputation is justly deserved and is based on performance. In addition, however, our present attitude has been shaped by critical formulations provoked by the general concern over the misery and distortion that accompanied industrialization. But because it initially took place in capitalist countries, industrialization is usually treated as a capitalist phenomenon—particularly when it comes to consideration of the evils of the process. Partly for this reason, the existence of environmental disruption in socialist and communist societies has been largely ignored.

Disturbed by the tumult that industrialization brought in its wake, those who sought to temper and ameliorate its more severe effects have searched for corrective proposals. Generally they have sought to reemphasize the importance of man and subordinate the machine. A good example of this kind of effort is the work of the various utopian theorists of the early nineteenth century. Subsequently, certain ideas of Karl Marx and Friedrich Engels reflected a similar reaction. They, like other writers of the era, were primarily concerned with questions of man's alienation from his work and the growth of social chaos. Hence at the time relatively little attention was paid by social theorists to questions of conservation and the preservation of nature.

Undoubtedly if nineteenth-century theorists had been transplanted to the twentieth century, almost all of them would have devoted a good deal of their work to environmental matters. While some environmental conditions in the nineteenth century were in many ways more serious than they are now, the main priority of the era was the resolution of some of the more pressing social injustices. Moreover, technology was such that few causes or cures for environmental disruption were known at the time. As a result, there was little that a fervent conservationist could do even if he wanted to, except perhaps to close down the polluters or flee to Walden Pond. The increasing concern devoted to conservation in the latter part of the twentieth century does not mean that all our other social problems have suddenly been solved. Rather environmental disruption has been aggravated by the growth of population and technology so that the potential for trouble has increased enormously. More and more observers have come to feel that today the potential hazards of environmental disruption are as serious as those of social disruption.

Radical Theorists' Views of the Environment
Given the dominant emphasis on social and political
concerns during the nineteenth and early twentieth
centuries, what if anything did radical theorists such as
Marx, Engels, and Lenin have to say about pollution?
It is hard to provide an objective answer. The three were
so prolific that if the scholar digs deep enough, he is
bound to find something that appears relevant to today's
concerns. The only difficulty is determining exactly how
relevant such citations were at the time. After all, when
they wrote the sharing of social wealth and the fruits of
economic development were the matters that were most
important to them.

Marx Of the three, Marx concerned himself the least
with the environment. At one point he does say, "Even a
whole society, a nation, or even all simultaneously existing societies taken together, are not the owners of the
globe. They are only its possessors, its usufructuaries,
and like *boni patres familias*, they must hand it down to
succeeding generations in an improved condition."
(Marx 1959b, p. 757). He also noted that if the soil is
properly treated, its productivity increases all the time
(Marx 1959b, p. 762). But all of this is offered in the
context of a long discussion of how land rent is determined and has little to do with ecology or conservation. In
other words, Marx is saying that if the earth is not maintained, rent values will fall. This would hardly qualify
him for membership in the Sierra Club or Friends of the
Earth.

Except for such isolated and out-of-context statements,

there is not much else that is relevant. It is true that Marx devoted page after page to the onerous working conditions in the factories of his time. He repeatedly attacked the noise, dust, heat, and smell that surrounded the worker and often his family (Marx 1959a, pp. 246, 425, 664). He was also highly critical of the lack of proper ventilation and sewage facilities. But generally these complaints were raised in the context of how the factory worker was oppressed, and not as a general comment on the environment. Marx also devoted some attention to recycling but mainly in relation to its impact on the prices of raw materials (Marx 1959b, pp. 79, 100). He noted the difference between reusing and preventing waste, but again only as it affected supply and demand conditions, not the environment (Marx 1959b, pp. 101–102).

Engels By contrast, Engels appears to have been surprisingly sensitive to questions of environmental quality. In what can be regarded as a most perceptive statement for either the nineteenth or twentieth centuries, Engels recognized the hazards man faces when he attempts to improve on nature. As he put it,

Let us not, however, be very hopeful about our human conquest over nature. For each such victory, nature manages to take her revenge. Each of these victories, it is true, has in the first place the consequences on which we counted. But in the second and third phase (secondary and tertiary effects) there are quite different, unforeseen effects which only too often cancel out the significance of the first.

Just like a modern-day conservation buff, Engels went
on to point out that

The people who in Mesopotamia, Greece, Asia Minor,
and elsewhere destroyed the forests to obtain cultivable
land, never dreamed that they were laying the basis for
the present devastated condition of these countries, by
removing along with the forest, the collecting centres and
reservoirs of moisture. When, on the southern slopes of
the mountains, the Italians of the Alps used up the pine
forests so carefully cherished on the northern slopes, they
had no inkling that by doing so they were cutting at the
roots of the dairy industry in their region: they had still
less inkling that by doing so they were thereby depriving
their mountain springs of water for the greater part of
the year, with the effect that these would be able to pour
still more furious flood torrents on the plains during the
rainy seasons. . . . Thus at every step we are reminded
that we by no means rule over nature like a conqueror
over a foreign people, like someone standing outside
nature—but that we, with flesh, blood and brain, belong
to nature, and exist in its midst, and that our mastery of
it consists in the fact that we have the advantage over all
other beings of being able to know and correctly apply
its laws. (Engels 1940, pp. 291–292; Engels 1955, pp.
140–141).

Even if Engels then goes on to suggest that revamping
nature is not so hard after all, his warnings are remark-
ably enlightened for the nineteenth century, and they are
particularly relevant today, as we shall see in Chapter 7.
Significantly, his remarks on the environment pertain
to society as a whole, not just to the private businessman.
Lenin Unlike his ideological mentors, Lenin could not
restrict himself to ideology. As the ruler of a country, he

left his revolutionary tower and took up the challenge of administering and meeting day-to-day payrolls and operations. Consequently, he had a greater opportunity and need to involve himself with what we would today regard as environmental questions. But since Lenin's main concern was economic development, not the environment, there has been a tendency to attribute beliefs posthumously to Lenin that in no way entered his mind (Zile 1970a, p. 5). As often as not, Lenin sought to achieve his goal of economic development by using rather than preserving Russia's rich natural endowment. And when he did display proper concerns about the environment, as laudable as they may have been, they were not as advanced as those of someone like Theodore Roosevelt.

Indicative of the fact that Lenin was a latecomer to questions of environmental disruption is the absence, prior to the mid 1960s, of reference to Lenin in discussions of environmental problems (Kolbasov 1965, pp. 1–50; Zile 1970a, p. 2). With the approach of the centennial of Lenin's birth on April 23, 1970, and the worldwide concern about environmental disruption, it was inevitable that Soviet scholars would dig back into Lenin's collective works and acts to see what might be related. In the welter of romanticism surrounding Lenin, the temptation has been to bring in too much.

Perhaps the best analysis of Lenin on the environment is Zigurds Zile's article "Lenin's Contribution to Law: The Case of Protection and Preservation of the Natural

Environment" (Zile 1970a). Zile pointed out that, indeed, many laws of a conservationist nature were passed during Lenin's administration.

The first caution to note, however, is that with the passage of time it becomes increasingly hard to show whether Lenin was the prime mover or just the automatic approver. As Chairman of the Council of People's Commissars, the chief administrative officer in the country, and the President of the Council of Labor and Defense, Lenin's signature was affixed to all laws and decrees. Today this is taken as proof that these were Lenin's laws. Undoubtedly some of them were; but logically the bulk of them originated at lower levels (Zile 1970a, pp. 11, 21, 22, 24). Yet there is also danger of overreacting to Soviet puffing about the contribution of Lenin. Regardless of who the initiator was, and even though implementation was generally neglected, the fact remains that many far-sighted laws on conservation were passed in the formative and hectic years of a new society. For several years after Lenin's death, subsequent legislation was in no way so conservation oriented.

Among some of the major laws and actions that are now printed out whenever anyone pushes the button "Lenin the conservationist" are the following:

1. The Decree about Land and the Socialization of Land of November 8 (October 26, old calendar), 1917, which claimed all of Russia's natural resources for the state (Kazantsev 1967, p. 6; Zile 1970a, p. 9).

2. Socialist Land Use, February 14, 1919, which among other things called for the protection of land resources

and their irrigation and drainage (Kazantsev 1967, pp. 6–7).

3. Decree about Forests, May 27, 1918, along with supplementary laws in 1920 and 1923 which attempted to protect the forests, parks, and gardens from the ravages of the world and civil wars and the exploitation of the cold and hungry populace (Kazantsev 1967, p. 8; Zile 1970a, p. 11).

4. A spring 1919 law that established a Forest and Natural Wild Life Preserve at Astrakhan (Kazantsev 1967, p. 11; Zile 1970a, pp. 22–23).

5. The February 20, 1919, decree of the Supreme Council of the National Economy, "On the Central Committee for Water Conservation," signed by Lenin. This law was intended to protect reservoirs from pollution by sewage and to provide for the prevention of damage from the sewage of factories, municipalities, and state installations. It also authorized the establishment of a central water-protection agency (Kolbasov 1965, p. 211). (A fuller listing of the laws passed under Lenin's administration is presented in Appendix A.)

In addition to the laws, however, modern-day conservationists in the USSR take pleasure in recounting anecdotes to illustrate Lenin's concern for the environment. Their favorite story is one originally told by Lenin's chauffeur. It seems that Lenin personally discovered a group of workers from the Bogatir' Factory cutting trees in Sokol'niki Park and ordered them to stop (Kazantsev 1965, p. 7; Zile 1970a, p. 15). A directive barring timber cutting in suburban forests within thirty versts (about twenty miles) of Moscow was then ordered by Lenin. On this count, Lenin certainly outshines George Washington.

The second caution about regarding Lenin as a great conservationist is that despite these environmentally oriented laws and regulations, countervailing laws and regulations were also issued frequently, and Lenin also signed these. To provide needed supplies of fuel and timber for export, Lenin readily approved the wholesale destruction of many forests (Zile 1970a, p. 13). In other words, as long as conservation did not stand in the way of economic need, it was a laudable goal. Unfortunately, the definition of economic need is not rigidly fixed, and therefore it was shaped to meet the convenience of the parties involved even if it came at the expense of nature. Furthermore, in an era of political and economic unrest, conservation regulations, along with almost all laws at the time, were ignored as often as they were honored.

The Bourgeois Line on the Environment

Since Marx, Engels, and Lenin devoted so little attention to an analysis of the responsibility of private industry for the disruption of the environment, it is probable that non-Marxist economists are responsible for the pervasiveness of the attitude that environmental disruption stems almost solely from privately owned industry. Of course, to the extent that conservationists looked out their windows and saw that it was private industry that was polluting, there was no need for any fancy elaboration of why it was that their intuition was widely accepted by leading economic theorists. Moreover, most

Western conservationists have had no experience with state-owned industry, and so they had no reason to believe that state-owned industry would abuse the environment.

The English economist A. C. Pigou is largely credited with the first systematic explanation of why it is that private industry tends to engage in such antisocial acts as pollution. The message in his book *Wealth and Welfare* is as relevant today as it was in 1912 (Pigou 1912, p. 159). As he explained it, industry need have no concern about what kinds of effluent it discharges because any damages created in the process are regarded as social costs or social products, whereas the manufacturer has to concern himself only with private costs or private net products. As an illustration of the divergence between social and private net products he provided the example of factory "smoke in large towns [which] inflicts a heavy loss on the community in respect of health, of injury to buildings and vegetables, of expenses of washing clothes and cleaning rooms, of expenses for the provision of extra artificial light . . .," which comes about because there is no way to force private polluters to bear the social cost of their operations (Pigou 1912, p. 159).

Pigou returned to this theme in a subsequent book *A Study in Public Finance* (Pigou 1928, p. 115). In his words,

. . . the value of the marginal social net product falls short of the value of the marginal private net product when resources yield, besides the commodity which is sold and paid for, a discommodity for which those on whom it is inflicted are unable to exact compensation.

Elaborating, he goes on to say,

Thus incidental uncharged disservices are rendered to
third parties when the owner of a site in the residential
quarter of a city builds a factory there and so destroys a
great part of the amenities of neighbouring sites; or when
he invests resources in erecting in a crowded centre,
buildings, which, by contracting the air space and the
playing room of the neighbourhood, injure the health
and efficiency of the families living there.

Hence, Pigou was one of the first to define the dis-
crepancy between social and private benefit and social
and private cost. In the years that have followed, this
social cost has been referred to by some as a problem in
externalities or as neighborhood or spillover effects.
Pigou's pointing to private benefit at the expense of social
costs for society at large undoubtedly helped to establish
the notion that private (not public or state-owned)
interests were the source of such publicly shared un-
pleasantness as pollution.

A concrete example of external costs is the polluter's
passing on to society of a large portion of the damage he
generates. Thus, in order to hold their private costs
down, electric plant operators normally try to avoid the
expenditures necessary to treat their chimney effluent
before it leaves the stacks. Instead they prefer to scatter
their airborne waste around the countryside so that those
downwind end up bearing the cost. Similarly, chemical
factories and paper pulp mills have traditionally utilized
nearby streams for the disposal of their raw wastes. This
obviates the need for in-plant treatment equipment,
which may sometimes necessitate expenditures of as much

as 10 percent of the basic capital cost. Since such costs are pushed off onto society at large, the polluter is able to avoid part of the real cost of his production. Normally this makes it possible for him to produce more and sell at a cheaper price than would be the case if the producer were forced to internalize and thereby bear his external costs.[1]

Pigou's ideas have become part of the mainstream of economic thought. Textbooks as well as specialized studies acknowledge the shortcomings that externalities impose on the market. For example, Paul Samuelson in his eighth edition of *Economics* refers to the external diseconomies or externalities almost exclusively in terms of private business, and such special studies as those of Karl William Kapp and E. J. Mishan are predicated on the same concepts (Samuelson 1970, pp. 453, 792; Goldman 1967, p. 82; Mishan 1967, p. 29). Even a laissez-faire economist such as Milton Friedman accepts the need for government intervention and regulation in order to prevent pollution (Goldman 1967, p. 80). In Friedman's words, "Strictly voluntary exchange is impossible . . . when actions of individuals have effects on other individuals for which it is not feasible to charge or recompense them. This is the problem of 'neighborhood effects.' An obvious example is the pollution of a stream."

Given the virtually unanimous agreement that private firms are unable to prevent pollution if left to themselves, it seems logical to call for government intervention. But not all economists stop there. Since private industry

[1] J. M. Buchanan has found an exception to the general rule (Buchanan 1969, p. 174).

cannot solve the problem when left to its own resources, some economists have gone on to conclude that public ownership of the means of production will prevent pollution. Typical of such reasoning is the argument of Oscar Lange, who stated that a "feature which distinguishes a socialist economy from one based on private enterprise is the *comprehensiveness* of the items entering into the [socialist] price system." In the private enterprise system, however, Lange asserts that no such comprehensiveness exists. Citing Pigou, he says,

> Professor Pigou has shown there is frequently a divergence between the private cost borne by an entrepreneur and the social cost of production. Into the cost account of the private entrepreneur only those items enter for which he has to pay a price, while such items as the maintenance of the unemployed created when he discharges workers, the provision for the victims of occupational diseases and industrial accidents, etc. [pollution—M.I.G.] do not enter. . . . (Lange and Taylor 1938, pp. 103–104.)

There was no doubt in Lange's mind, however, that such costs would be included under socialism.

Soviet Law—Perfection Itself

To a limited extent at least, present Soviet policy makers reflect the trend of such thinking. The influence of the non-Marxist school is probably minimal, but Marx, Engels, and Lenin undoubtedly have had an impact. If nothing else, their view of the superiority of the public sector over the private sector in all areas has affected current attitudes toward the superiority of the Soviet approach to pollution control. A speech by B. V. Petrovsky, the head of the Ministry of Public Health of the USSR at a 1968 session of the Supreme Soviet is illustrative.

"Problems related to air and water pollution are being discussed in a host of capitalist countries. But the capitalist system, by its very essence, is incapable of taking radical measures to ensure efficient conservation of nature." In contrast he asserts that "In the Soviet Union questions of protecting the environment from pollution by industrial wastes occupy the center of the Party's and government's attention. It is forbidden to put industrial projects into operation if the construction of purification installations has not been completed." (*Prav* 6/26/68; *CDSP* 7/17/68, p. 16.) Similarly, "The Soviet Union has been the first country in the world to set maximum permissible concentrations of harmful substances in the air of populated areas."

By implication Petrovsky suggests that the Soviet Union has acted to prevent or eliminate pollution. Given such an intellectual climate, it must be somewhat embarrassing for Nikolai Popov, an editor of *Soviet Life*, to have to ask, "Why in a socialist country whose constitution explicitly says the public interest may not be ignored with impunity, are industry executives permitted to break the laws protecting nature?" (*SL* 8/66, p. 3.) Similarly, Academician Innokenty Gerasimov asks, "What is it in our society with its consistent progress in all spheres of life, that interferes with a rapid advance in such an extremely important field as the rational exploitation of nature?" (Gerasimov 1969, p. 75.)

Certainly the existence of environmental disruption in

the USSR cannot be explained by a lack of legal concern with pollution. (For a listing of Soviet laws dealing with the environment, see Appendix A.) If anything, from the time of Lenin to the present there has been a heavy reliance on legal restraint and good intentions. Regrettably, Soviet emphasis on the legal formalities has generated a form of self-deception. This appears to be an instance where some authorities have been lulled into believing that respect for Soviet authority is such that the mere passage of highly desirable laws is all that is necessary to induce compliance. Such self-deception is, of course, not always limited to socialist countries such as the USSR. The danger in such situations is that exaggerating the superiority of one's system frequently leads to overreaching and overcommitment. If the goals are set too high, those who are supposed to implement them become cynical so that similar efforts in the future may be discredited before their feasibility can be evaluated. By setting unattainable goals after the 1917 revolution when enforcement powers were too weak or lacking altogether, Lenin bred into the system of environmental control a certain contempt that persists today.

For example, it may be true, as the Minister of Public Health Petrovsky asserted, that the Soviet Union was the first country in the world to set maximum permissible concentrations of harmful substances in the air. It also was one of the first to set limits on the discharge of various types of water effluent. In both cases the maximum norms were generally lower than those established in other countries of the world (see Table 1.1). For example,

Table 1.1 Maximum Levels of Concentration of Selected Toxic Substances in Atmosphere near Populated Areas

Chemical	Maximum Concentration (milligrams per cubic meter)	
	Maximum Level	Daily Average
Acetate	0.10	0.10
Acetone	0.35	0.35
Acetophenon	0.003	0.003
Acrolein	0.30	0.10
Aniline	0.05	0.03
Arsenic	—	0.01
Benzol	2.4	0.80
Butylen	3.	3.
Carbon	0.15	0.05
Carbon disulfide	0.03	0.01
Carbon monoxide	6.	1.
Chlorine	0.10	0.03
Chromium	0.0015	0.0015
Dust	0.5	0.15
Dynel	0.01	0.01
Fluoride compounds	0.03	0.01
Formaldehyde	0.03	0.012
Hydrochloric acid	0.05	0.015
Hydrogen sulfide	0.008	0.008
Lead	—	0.0007
Lead sulfide	—	0.0017
Manganese and its compounds	—	0.01
Mercury (metallic)	—	0.0003
Methanol	1.5	0.5
Nitric acid	0.01	—
Nitrogen oxide	0.3	0.1
Phenylic acid (carbolic)	0.01	0.01
Phosphoric anhydride	0.15	0.05
Propylene	3.	3.
Styrol	0.003	0.003
Sulfur anhydride	0.5	0.15
Sulfuric acid	0.3	0.1

Source: Sokolovskii et al. 1965, pp. 92–93.

Table 1.2 General Requirements for the Composition and Properties of Drinking and Non-Drinking Water in the USSR

	Water Used for	
Indices	Central and Noncentral Mains, Supply and Production Processes in the Food Industry	Swimming, Sport, Recreation, and Bodies of Water in Residential Areas
Matter in suspension	Matter in suspension not to exceed:	
	0.25 mg/liter	0.75 mg/liter
	Water with more than 30 mg/liter of natural minerals at average water level may exceed limits for matter in suspension by 5 percent. It is forbidden to release suspensions with an inflow rate of more than 0.4 mm/sec in flowing water or 0.2 mm/sec in standing water	
Floating matter	No filming, oil slicks, or other accretions should be detectable on the surface of the water	
Smell and taste	The water may not assume smells and tastes more intensive than 2nd Degree* in	
	Untreated state or after chlorination	Untreated state
	The water may not affect the smell or taste of the flesh of fish	
Color	Should not be detectable in a layer of	
	20 cm	10 cm
Temperature	The winter temperature may not exceed the maximum summer temperature by more than 3° C after release of waste matter	
Reaction	pH 6.5 to 8.5	
Mineral content	Undissolved remnant not to exceed 1,000 mg/liter, of which	Standard in accordance with factors under the heading "Smell and taste"
	chlorides—350 mg/liter sulphates—500 mg/liter	
O_2 in solution	At least 4 mg/liter regardless of season	

Biochemical O_2 consumption	At 20° C may not exceed	
	3 mg/liter	6 mg/liter
Pathogens	May contain no pathogens. Waste containing pathogens must be disinfected after pre-cleaning. The methods of disinfection and pre-cleaning (mechanical and biological) must in every case comply with the regulations of the State Sanitary Inspectorate	
Contaminants	May not occur in concentrations directly or indirectly endangering public heatlth	

* Russian text does not specify scale.
Source: Eva Maria Kraus, "Combatting Water Pollution in the USSR," *Review of Soviet Medical Sciences*, Vol. 2, No. 4, 1965, p. 56.

according to our Clean Air Act of 1970, maximum levels of carbon monoxide in the United States are not to exceed 10 milligrams per cubic meter in an eight-hour period and 40 milligrams per cubic meter in a one-hour period (*Federal Register*, vol. 36, no. 84, p. 8187). The maximum level in the USSR according to Table 1.1 is only 6 milligrams per cubic meter. Yet when asked if such stringent standards were enforced, one candid Soviet authority told me, "No, enforcement of such standards would cripple all industrial production and municipal life." Why then were such laws instituted? "As a sign of what a socialist system can do." This may be good for external public relations, but it is likely to have the opposite effect on those who carry out the laws (*EG* 9/18/65, p. 15; *Prav* 7/6/70, p. 5).

The violations and the weak enforcement of the laws support the theory of legal impotence (Sushkov 1969, p. 6). Unless Soviet conservation authorities are addressing a foreign audience, virtually all of them agree that Soviet laws for the protection of nature are poorly observed (*Prav* 8/31/69, p. 2; *Prav Vos* 11/1/68, p. 3; *Sov Est* 8/19/70, p. 2; Ianshin 1965, p. 57). In the words of Professor D. Armand, the respected ecologist, "It is against the law in the Soviet Union for any factory to release emissions into the air without the installation of appropriate filters. Factories are obligated to install them. However, the existing law is badly enforced." (Armand 1966, p. 150; *Sov Lit* 5/15/71, p. 3.)

The Post-Lenin Legislation
For more than three decades after Lenin's death in 1924, slight attention was paid to preserving the country's natural resources. There was little enforcement of existing laws and almost no enactment of new laws. All eyes and bodies were focused on economic development, the fulfillment of the five-year plans, and the winning of World War II. Stalin did not allow for much else. With the exception of a prerevolutionary law passed in 1914–1915 by the Moscow regional municipal councils (Zemstvo), which set up a zone of Sanitary Protection for the Moskovorets (the Moscow water supply agency), there was little else to work with beyond the laws passed under Lenin's administration (Kolbasov 1965, pp. 104, 164). But ecological interests were not important to the Soviet leaders of the day. Among the few laws passed in the three decades following Lenin's death that related to conservation were a resolution about fishing, a law setting up sanitary protection zones around drinking reservoirs,

and a regulation for health resorts. There was also Stalin's massive scheme for the creation of a shelter belt of trees announced in October 1948 (Shabad 1951, p. 58). Eight huge tree belts were to be planted in strategic locations across the country, particularly in areas affected by drought and wind erosion. The total length of these belts was to be 3,500 miles. But with the shelter belt plan as well as with the other laws, Lenin's included, very little attention was paid to their implementation, even when personal health was involved.

The one substantive and meaningful innovation in the field of environmental law enacted during the Stalin era occurred in the field of air pollution. In 1949 the Council of Ministers of the USSR passed what was probably their first resolution on air pollution. Among other things, the resolution provided for the establishment of the Chief Administration for Sanitary Epidemiological Supervision. But of all the actions during the whole period, the most meaningful, as we shall see in Chapter 4, was the improvement which took place in the quality of urban air, particularly in Moscow.

The air pollution resolution in 1949 was the first hint that environmental questions would eventually have to be taken seriously. Because of the rapid rate of industrial growth, such a transformation was only a matter of time. Nevertheless, almost another decade passed before legislative concern took on serious proportions (Kolbasov 1965, p. 213). Even then, however, the typical response to repeated assaults on the country's natural resources was to pass another law, a phenomenon not unique to the

USSR. As a result, beginning on June 7, 1957, with the passage of a law "For the Protection of Nature" in the Republic of Estonia, until Khrushchev's ouster in the fall of 1964, enough laws on the environment were passed to satisfy the most insatiable lawyer. By March 26, 1963, all fifteen Soviet republics had laws "For the Protection of Nature" on their books (Kazantsev 1967, pp. 16, 37). Only a few of the more important other laws will be listed here, but if one were to judge purely on the basis of the number of relevant laws and decrees that had been issued, it would be Khrushchev, "the unperson," not Lenin, who would be memorialized as the father-guardian of conservation in the USSR. According to a count made by Zigurds Zile, between 1956 and 1960 at least nineteen executive decrees on conservation and ten articles in a new criminal code devoted to conservation had been issued in the Russian Republic (RSFSR) alone (Zile 1970b, p. 5).

The law "For the Protection of Nature" adopted on October 27, 1960, by the Russian Republic is typical of those adopted in the other republics. (Its text is reprinted in Appendix B.) It spells out which resources are subject to control: land, mineral resources, water, forests, typical and rare landscapes, resort and green-belt areas, wild animal life, and the atmosphere. All factories are ordered to provide themselves with purification equipment so that there will be no discharge of sewage that is not treated and purified.

For some reason, these general laws for the protection of nature were apparently not considered adequate. Supplementary ones on fishing, hunting, underground water supplies, mining, geological work, land drainage and irrigation, wind and water erosion, sewage, timber, green belts, and air were passed at the republic and All Union levels (Kazantsev 1967, pp. 13–14, 43–50, 54–55; Kolbasov 1965, p. 54; see Appendix A). At least one specialist explained the need for this redundancy by acknowledging that the basic law for the preservation of nature in the RSFSR was being "unsatisfactorily fulfilled" (Kazantsev 1967, p. 54).

The plethora of Soviet laws passed in the late 1950s and in the 1960s demonstrates that the accepted approach to the solution of environmental disruption in the USSR seemed to be, if the first law does not work, pass another one. (Again, this is not a practice peculiar to the USSR.) Often it seemed as if the second law differed only slightly, if at all, from the original one already on the statute books. On May 9, 1960, for example, a law was issued by the Council of Ministers of the Russian Republic guaranteeing the preservation of Lake Baikal (Trofiuk and Gerasimov 1965, p. 58). Specific and reasonable measures were spelled out which if fulfilled would have indeed preserved the uniqueness of the lake. Unfortunately, they were not fulfilled. In lieu of concrete actions, another law preserving Lake Baikal was passed on February 7, 1969. This time, however, it was passed by the Council of Ministers of the USSR (*Iz* 2/8/69, p. 2). Since there is little difference between the two laws, it is unlikely that the second law will be much more effective that the first.

The passage of a second law covering the same situation

is generally a good clue that the situation has not been helped much by the first law. Among other duplicate sets of laws are

1. A March 3, 1962, and a February 26, 1969, law spelling out measures for the protection of the Black Sea Coast.

2. A law of November 27, 1959, and a similar one of October 26, 1965, on the protection of nature in Latvia.

3. A law of April 22, 1960, and December 11, 1970, regulating water resources. (See Appendix C for the complete text of the 1970 law.) Over the course of the decade they changed the title so that the 1970 law is called "The Principal Water Legislation of the USSR and the Union Republics," whereas the 1960 law was titled "On Measures to Regularize the Use and Increase the Protection of the Water Resources of the USSR." Apparently little else of substance was altered.

Skepticism about the effectiveness of the new laws persists. Enforcement is still weak. The best evidence of this is the contrast between the number of pollution violations that take place and the number of legal suits that are instituted or penalties that are collected. It is charged that often the control commission makes no effort to enforce the law (*EG* 8/18/65, No. 33, p. 15). Even if the violators are brought into court, there is no guarantee of remedy since, as T. S. Sushkov, the President of the Commission for the Protection of Nature of the Supreme Soviet of the RSFSR, put it, "The courts have been too lenient with polluters." (Sushkov 1969, p. 5.) He reports that courts of the RSFSR convicted only five people in 1963 and five in 1965. The figure for 1966 was ten, and in 1967 it was twelve.[2]

Some conservationists even question the validity of the slight increase in prosecutions that have taken place in the USSR. At a conference of Soviet specialists on the legal problems in conservation, the complaint was made, à la Ralph Nader, that the main thrust of conservation law in the USSR was directed against the wrong targets (*Priroda* 1/70, p. 119). So far it is the poacher (that is, individual citizens) that are being harassed, while government institutions (factories and municipalities) are frequently left to themselves. The real polluters for the most part are not rigorously regulated or penalized. For that matter, this author is unaware of any legal action brought against a municipality or noneconomic state facility (army base, hospital) in the USSR. Yet it is the government institutions (municipal, service, manufacturing, and agricultural) that are responsible for damage hundreds of times more destructive than that of the poachers. As will be discussed at greater length in the next chapter, to some extent this is due to the absence of an effective form of federalism in the USSR. Thus there is no system of checks and balances and no pressure by the central government on the local or state

[2] Convictions were also rare in the United States until the 1970's. After the mercury scare of 1970 and the simultaneous discovery of the Refuse Act of 1899, enforcement and conviction in the United States have been pursued vigorously. Even though there is a similar Russian law that requires a factory to obtain a permit and renew it every three years before it can discharge anything into a water course, like enforcement of all Soviet pollution laws, no parallel enthusiasm for this law has been noted in the USSR (Kolbasov 1965, p. 223).

(republic) governments to treat their environmental disruption. To a larger degree than in the United States, the overseers are also the overseen. It is unrealistic to expect polluters to punish themselves for their offenses. That is exactly what must happen in a society where the state is also the polluter and the manufacturer.

On those infrequent occasions when industrial polluters in the USSR are brought to court, there is no guarantee that the court's injunctions will be enforced. On one occasion the manager of a handicraft plant on the Volga was found guilty of dumping wood chips into the river, resulting in the death of a large number of fish. After he was ordered to pay a fine from his own pocket, he appealed through the nonjudicial channels of the local and regional Communist party organizations. Instead of supporting the verdict, these organizations attacked the state inspection service for bringing the case up in the first place (*Sov Ros* 2/15/70, p. 2). Only after publicity was focused on this illegal procedure was the decree reinstituted and the fine paid (*Sov Ros* 4/26/70, p. 2).

The Power of a Fine
Despite the propensity of all governments to rely on fines to implement their laws, fines are generally not much of a deterrent. This is particularly true in the USSR. According to Soviet law, there are two types of violations. Although it is not always clear where one ends and the other begins, one category is called minor violations and is subject to administrative penalties. More serious violations are sometimes treated as criminal acts (Kazant-

sev 1967, pp. 57–58). The administrative penalties are
not very intimidating. The resolution of the Council of
Ministers of the RSFSR of February 18, 1963, provides
that the pollution of the air or water can result in a fine
of up to 50 rubles (about $56) for the person primarily
responsible and up to 10 rubles (about $11) for others.
For particularly serious violations, the fines can be increased to 100 rubles (about $111) for the person in
charge and 50 rubles for the others involved. In most
areas the 100-ruble fine is the norm; in a few republics
the usual fee is as low as 5 to 50 rubles. (Vitt 1970, p. 78;
Sov Mol 6/11/69, p. 2.) It is hard to see how such fines
could have any impact.

When a determined stand is taken and a stiff fine is
levied, the fine may still not serve as a deterrent even
under the criminal code. In most cases the fine is not
taken out of the plant director's pocket but out of the
enterprise fund (Vitt 1970, p. 78). Under these circumstances the plant manager hardly feels any personal loss.
Professor Armand, one of the Soviet Union's most
respected ecologists, reports that funds to pay the fines
were frequently included in the financial plan of the
enterprise at the beginning of the year (Armand 1966,
p. 71; Nagibina 1961, p. 295). Under such circumstances
the plant manager may actually come out ahead if he
pays the fine and refuses to halt production so that
pollution control facilities can be installed or repaired.
By continuing with production and pollution he will
stand a better chance of meeting his production targets.
This in turn will entitle him to high salary premiums.

(See Chapter 2.) Even though the manager may come out ahead, the same cannot be said of the polluted water body.

Compounding the situation, local officials until January 1, 1962, and in several cases thereafter actually had a vested interest in seeing that conditions were not improved. As long as the fees were collected and turned over to local government bodies, they stood to benefit from the levying of such fines (Armand 1966, p. 71). In some instances the local soviets counted on collecting these funds in order to finance the construction of local facilities like hospitals and other public facilities. Consequently anything that was done to eliminate pollution also meant that the local government authorities would lose access to these supplemental funds. Furthermore, the fines did not really cost the local firms much since usually the necessary funds were allocated centrally by the ministries specifically for this purpose (Armand 1966, p. 71). Because of the obvious conflict of interest, this system was finally abolished though there are some authorities who want to return to it (*Iz* 3/25/71, p. 4; *CDSP* 4/27/71, p. 23).

Under administrative procedures, government authorities also have the right to close down polluting enterprises if proper treatment equipment is not installed. Just as in the United States, this power is used sparingly. Professor Armand complained that, as of 1963, this power had never been used despite some particularly gross violations. According to Professor Armand, Gosvodkhoz (the State Water Economic Administration) of

the RSFSR, which has this authority, has not exercised
it. "The reason for such indulgence," Armand explained,
"is that the product of the factory is very necessary and
the plan cannot be disrupted." (Armand 1966, p. 69.)
We shall meet this problem again in Chapter 2.

Subsequently, as the magnitude of environmental
disruption continued to grow, there developed an increased need to deal with some of the more blatant
violators of the law. During the latter part of the 1960s,
some factories such as the Voskresensk Chemical Plant
were reportedly closed as well as some washing units at a
motor vehicle enterprise (*LG* 6/13/70, p. 11). Similarly,
the sewage outlet of the Novogork'y Oil Refinery on the
Volga was plugged up until repairs were made, and the
lead paint shop of the Krasnogorsk Chemical Plant was
closed down because of the particularly hazardous forms
of air pollution it was creating (*Trud* 12/25/70, p. 2; *LG*
8/9/67, p. 10; *CDSP* 9/6/67, p. 10; Nuttonson 1969b,
p. 2). Apparently the law is enforced this way only when
the physical health of Soviet citizens has been directly
affected.

In theory, at least, criminal violations are supposed to
be much more severe. According to Article 223 of the
Criminal Code of the RSFSR, anyone who threatens
human health or agricultural products or fish by polluting faces a fine of up to 300 rubles (about $333) or correc-
tive work of up to one year[3] (Kazantsev 1967, p. 59;
Trud 2/8/67, p. 2; Powell 1971, p. 629). In Turkmenistan, a law was passed on September 12, 1964, which

[3] Rarely has this meant jail, however.

increased the scale of penalties to up to three years of
"deprivation of freedom" or up to one year of corrective
labor or a fine up to 500 rubles (Kazantsev 1967, p. 59).
For the most part, however, the fine is levied as a
percentage of the violator's annual wage, usually 10 or
20 percent (*CDSP* 3/31/70, p. 17; *Sov Est* 4/26/70, p. 2).

Even the careful reader may be forgiven at this point
for feeling a bit bewildered. The welter of different levels
of jurisdiction and their different provisions have inevitably led to confusion, contradiction, and minor chaos.
Several legal authorities have called for a codification of
all the legislation on the environment which is presently
scattered in widely unrelated laws (Kolbasov 1965, p. 29;
Prav 3/19/71, p. 3; Lesnikova 1970, p. 119). Instead of
solving the problem, the passage of law after law has
apparently only intensified it. As in the United States,
where different state laws often call for different standards, which in turn frequently conflict with federal law,
so republic laws in the USSR may often differ from one
another. For example, most of the Soviet republics
classify deliberate pollution of the air as a criminal
offense, whereas the republics of Uzbekistan, Georgia,
Azerbaijan, Lithuania, Kirgizia, and Estonia do not
(*CDSP* 1/17/68, p. 29; *Iz* 12/29/67, p. 3).

Just as the laws overlap and contradict, so do the enforcement and administering agencies. Again, as in other
countries of the world including the United States, one
can never be certain who is reponsible for what. In fact,
with the formation in 1970 of the Environmental Protection Agency in the Office of the President, there is more

coordination in the United States than in the Soviet
Union. Among the agencies responsible for implementing
pollution control in the USSR are the Council of Ministers of the USSR as well as the Councils of Ministers of
the various republics, Gosplan (State Planning Office)
at the All Union and republic level, the Ministry of
Agriculture, the Ministry of Health, Committees for the
Protection of Nature in several of the republics, the
Ministry of Land Reclamation and Water Management,
the Ministry of Fisheries, and the Ministry of Electrical
Energy (*Sov Mol* 6/11/69, p. 2; *Sot In* 6/20/70, p. 2; *EG*
8/18/65, No. 33, p. 15; Kolbasov 1965, pp. 14, 29, 33;
Armand 1966, p. 29; *Trud* 11/12/66, p. 2; Kazantsev
1967, pp. 20–21).

  Because of the uncertain jurisdiction and inadequacy of
laws that seek to remedy pollution damage after the
plant has been built, there has been some movement
toward trying to *prevent* pollution by denying the plant
permission to operate until controls are installed and
working (*Sov Lat* 2/8/70, p. 4; *Sot In* 8/20/70, p. 3; *Trud*
2/18/67, p. 2). As we shall see, this rule has been broken
as readily as any of the others (Kolbasov 1965, pp. 226–
227). Theoretically, such a system should be effective
because each factory in the USSR is supposed to obtain
approval from the State Sanitary Inspectorate of the
Ministry of Health before any factory can open (*Prav*
6/21/65, p. 2; *CDSP* 7/7/65, p. 13; Kolbasov 1965, p. 226).
Moreover, laws have been passed which are supposed
to deny premiums to construction workers who complete
factory construction work on or ahead of time but who

fail to complete construction work on sanitary treatment facilities (*Sov Lat* 2/8/70, p. 4). Nonetheless, such laws again seem depressingly ineffective.

To those who believe in the sanctity of the law, such violations are hard to understand. But there are many who fail to understand that man does not live by law alone. Such people refuse to come to terms with economic realities. After much debate, the Soviet Union passed the basic water law of December 11, 1970, without requiring a fee for water use and discharge. No matter how cogent the reason for the introduction of such a fee, the economic rationale did not prevail. As explained to me by a Soviet economist in Moscow a few days after the passage of the law, the lawyers and political theorists had won out. Their argument was based partly on their adherence to Marxist ideology, which they interpreted as holding that the country's natural resources belong to all the people and therefore should be free. Representative of these attitudes is the following statement of O. S. Kolbasov, a leading environmental jurist, "The principle of using water resources permanently and without charge is inherent in the law of water use in the USSR. Natural reserves of water are always allotted for use without charge and in the overwhelming majority of cases for an indefinite period of time." (Kolbasov 1965, p. 28.)

For the time being, the lawyers and theorists have defeated the economists who argued that since water as well as air is scarce, there should be a fee for their use (*LG* 6/17/70, p. 11). Though they lost the debate, the economists will have to prevail sooner or later if there is ever to be an effective reduction of pollution. In the meantime, the real loser will probably be the environment. For an analysis of why it is that law alone cannot stem environmental disruption and why in fact economic forces in the Soviet Union can actually exacerbate the situation, we must turn to the next chapter.

2 The Economic and Political Propensity to Pollute— Reality

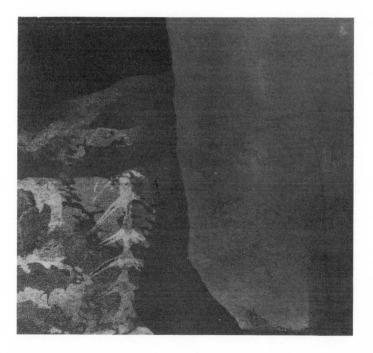

If it were only a question of ideology or hearts being in the right places, there would be no environmental disruption in the USSR. That should have been evident enough from Chapter 1. But contrary to all the doctrines and expectations of many social critics and reformers, state ownership of the means of production is no guarantee that social costs such as environmental disruption will be prevented (Lange and Taylor 1938, p. 103). The economic factors that have led to environmental disruption in the USSR, especially the role of external and social costs, will be examined here. Many of the underlying economic forces causing environmental disruption in the USSR also exist in the United States. Frequently the same economic phenomena are at work in all societies. Occasionally, however, there are certain pressures that are more likely to lead to environmental disruptions in socialist or communist economies than in mixed economies.

External Costs
Since Pigou emphasized the importance of external costs, one of the most common applications of the concept has been to explain why environmental disruption is so difficult to deal with. Environmental disruption will be inevitable in the Soviet Union as well as in the United States as long as the polluter is able to pass off a portion of the damage he generates onto society at large, and he will try to do this wherever possible. Thus Russian as well as American electric power plant operators will scatter their airborne waste around the countryside in

order to eliminate the need and cost of building air-treatment facilities on their plant sites.

As we have discovered in the United States, recognizing the principle of social cost is not the same as solving the problem. Even in the Ruhr Valley, where more than anywhere else an effort has been made to internalize the external costs of sewage disposal and treatment, only an approximate figure is used to estimate just how much damage is caused by the discharge of certain types of pollutants into the water. In other words, even when everyone agrees that the polluter should be made to bear his social costs, it is difficult to ascertain what these costs are.

The first obstacle is that there is little agreement as to just what is affected by environmental disruption. Those familiar with the debate are used to arguments about whether one should count only the primary costs of pollution or the secondary and tertiary effects as well. Even when this is decided, scientists and engineers are not always able to evaluate precisely the effects of certain forms of pollution. In some cases, the damage is too indirect or difficult to measure. Initially it may appear as if there have been no harmful effects. The damage becomes apparent only several years later. It took several decades before ecologists became aware of the harmful effects of DDT and mercury poisoning. To make internalization of social costs effective, economists should have precise data to work with during the initial stages of production. Still even a rough estimate is better than no estimate at all. The Genossenschaften, or cooperative water groups,

in the Ruhr make such estimates and then assess the
factories and municipalities for the approximate damage
they cause. This could serve as a model that might well
be adapted in many other water basins of the developed
world. Nonetheless, even though the Ruhr authorities
seem to have designed a workable system for coping with
water pollution, they have had considerably less success
in dealing with air pollution. This is largely because
responsibility for air pollution damage is much harder to
pinpoint and because methods of purifying the air do not
lend themselves to cooperative or joint treatment ven-
tures like those employed in water purification.

Externalities in a Socialist State

Because the state in a socialist society owns all the means
of production, sooner or later the state must bear all the
social costs. Thus, in theory, there should be no such
thing in the USSR as pushing social costs onto someone
else. Despite the difficulty of assigning responsibility to
each factory for its precise social costs, it should be in the
interest of the USSR to make each factory pay for the
social costs it generates. If each factory were held
accountable for both the direct and social costs of its
operations, much of the pollution would be treated within
the confines of the plant before it could be pushed onto
the population as a whole. Several Soviet economists,
including M. Loiter and I. Petrianov, have come to
accept this idea (Loiter 1967, p. 75; Petrianov 1969,
p. 76; "Otsenka prirodnykh resursov" 1968, p. 81). I shall
discuss some of the other proposals in Chapter 3, but

among the various suggestions is one by A. Vainshtein, who maintains that the cost of a project that has some major effect on the environment should include not just the direct costs of initial construction but the costs involved in cleaning up and restoring nature afterwards (*LG* 5/15/65, p. 2). In Vainshtein's view, this would provide a more complete analysis of just what the real costs are. Presumably, many otherwise acceptable projects would be rejected, and properly so.

In practice, however, almost no effort is made in the USSR to assign such social costs. Consequently, the Russian factory manager has no economic incentive to clean up his pollution himself. In fact, the difficulty of dealing with externalities is actually compounded in a country like the USSR. This is because the environmental authorities must contend not only with the uncertain consequences of technology and the lack of precise cost measurements that exist in nonsocialist countries but also with Marxist ideology. It would be very difficult ideologically for the Russians to impose a charge on air and water. The Russians are not necessarily the only ones who have to contend with such a sentiment. Many Americans also feel that the air and water are abundant resources that belong to everyone and should be free to all. In the USSR, however, this natural inclination has been reinforced by the ideological tenet that all natural resources, not just the air and water, belong to the state as a whole. Consequently, it seems to follow that since the state owns all the resources, it should provide them free to any user authorized by the state. In effect, state enterprises have been able to treat all the country's natural resources, including air and water, as if they were free goods. But even without complete profit maximization, when anything is free, there is a tendency to consume excess quantities of it without regard for future consequences. But like free love, there is a limit to how much of a country's natural wealth can be consumed. After a time, there is the risk of exhaustion.

In fact, many Russians by the 1960s began to express the fear that the country's richest resources would soon be exhausted. Especially disturbing were indications that much valuable ore and oil were being left in the mines or wells or were being discarded. Recovery rates appeared to be unusually low. Among other authorities, K. E. Gabyshev complained that many mines and oil wells in the country had only a 50 percent recovery rate (Gabyshev 1969, p. 18; *LG* 7/12/67, p. 11; *Iz* 8/8/70, p. 2; *Trud* 8/12/67, p. 2; *Sot In* 1/8/71, p. 2). The economist Iu. Sukhotin reported losses of 50 to 60 percent on the extraction of coal, oil, potassium, and natural gas (Sukhotin 1967, p. 88).

To this day the Russians still do not seem able to understand what the difficulty is. Apparently what happens is that after a Soviet mine operator has extracted the richest ore, his marginal costs and average variable costs begin to rise. As it takes more units of labor and machinery to extract one unit of ore and oil, the mine director begins to look for another, more easily exploited mine or oil deposit. This is a natural reaction since in the slang of the economist, "bygones are bygones," that is, the mine operator does not have to worry about recovering his old fixed costs. Economically they are treated as free goods. But since the new mining site and its mineral wealth are also treated as free goods, the mine director is tempted to move when his *marginal* costs at the old site (primarily labor, capital, and shipping) begin to exceed the cost figure at the new site. The cost at the new site in the USSR, however, includes only the *average labor and capital costs plus the average cost of moving the mining and drilling equipment* to the new site. The important point is that not only does the mine operator not have to worry about fixed costs (land) at the *old* site, but he also does not have to worry about fixed costs (land) at the *new* site. Furthermore, until July 1, 1967, even the geological exploration costs were not charged to the price of the raw materials (Feitel'man 1968, p. 10). After mid-1967, some but not all geological costs were included ("Ekonomicheskaia Otsenka prirodnyka resursov" 1969, p. 103). This contributes to the further understatement of expenses that will be incurred during any move to a new location.

In other economic systems, however, where the mining company has to buy the land or the mining rights, such a calculation would also include all the components of the *average cost of land per unit of output* at the *new* mining site. The inclusion of such a cost means that the cost of the oil or ore will reflect the present discounted value of the commodity in the future. Consequently the mine director in a non-Soviet economy will be less inclined to switch to a new site. Because the non-Soviet mine operator has to pay for new land costs, he will normally have a higher average total cost of production than if the land at the new site were treated as a free good. Therefore he will seek to extract a greater percentage of ore or oil from his original location. If the mine operator can sell the old mine, this may help defray the cost of the new mine site and therefore make it easier for him to move. However, the value of an abandoned underground or strip mine is usually not very high. In any case, for the Soviets, transfer to a new mine site becomes profitable at an earlier point. As a result, Soviet mine operators do not have to concern themselves with the intensive exploitation of their deposits. They simply move to a new site.

A by-product of this extensive system of mining has been that large quantities of salvageable raw materials are often discarded (*CDSP* 2/24/70, p. 11). Thus the low recovery rates have increased environmental disruption as potentially useful minerals have been added to the slag heap and oil wastes have been spewed into the country's water courses. As long as mine operators attempt to economize on their use of labor and machinery and not on their raw material deposits, this will be the result.

Interestingly enough, if mine operators and oil drillers in a given economic system are in the habit of renting rather than purchasing their mining and drilling sites, the tendency to move to a new site may be even higher than if the mine and its minerals are treated as free. This

comes about if the rents for the mines are not differentiated according to the richness of the deposit. Even
though the old site may have rich and untapped deposits,
if all rents are the same, the miner will be tempted to
move to a new site as soon as he finds that the total
average cost of the new site is lower than the marginal
cost per unit of output at the old site.

After numerous complaints about the waste of resources,
some Soviet economists began to call for the assignment
of prices to the country's raw materials. They sought to
do this by instituting rental fees for mining enterprises.
These proposals were implemented during the price
reform of July 1967. It is not entirely clear how these
rental charges will be applied. Presumably the pricing
authorities sought to design a system that would induce
miners and drillers to work their raw material deposits
more intensively so there would be less waste. However,
as just indicated, it is possible that unless the rents are
properly scaled, rental charges on mining land may
actually accelerate rather than retard mine abandonment. Theoretically there would be more intensive
working of the mines if the mine operators were required
to buy their mine sites, but this would be awkward
ideologically.

Ideology is not the only impediment to assigning the
proper cost to natural resources in the USSR. The state,
not the market, determines prices and costs in the USSR.
But because the task of setting prices is such an immense
undertaking, in many instances the State Committee on

Prices does a poor job. Given the proper circumstances, the centralization of price setting can be an important tool in protecting a country's natural resources. Thus the State Committee on Prices could conceivably decree that high prices should be assigned to those products that give rise to serious pollution in the course of their production. This would serve the dual purpose of providing funds for pollution control and of discouraging consumption of a particular product because of its high price. So far, however, there seems to be little evidence that the Russians have adopted such a policy in their pricing practices. Generally it seems fair to say that such questions have been ignored in the price-setting process. In fact, at least until July 1967, prices on raw materials were often set below the direct costs of production. V. K. Sitnin, the Minister of the State Committee on Prices, complained that prior to the price reforms of that year, many raw materials were mined at a loss (*EG* 2/68, No. 6, p. 10). Similarly, two economists, G. Borisovich and A. Vain, note that in the USSR rental payments in gas extraction are related directly rather than inversely to costs; thus the lower the costs, the lower the rent charge (Borisovich and Vain 1969, pp. 38–42). It is hard to see the justification for such a system. It probably accelerates the abandonment of high-cost wells and whatever recoverable oil they might contain. Just because a state organization has the power to set prices, it does not necessarily follow that it will use this power wisely or for the promotion of conservation.

Shortcomings in Cost Benefit Analysis

Even if a well-conceived and acceptable cost benefit analysis could be designed that would take into account social costs, there would still be obstacles to its successful implementation. Several of these obstacles exist in any society, but, again, some of them are more formidable in a socialist system.

General Difficulties Those who insist on wider use of cost benefit analysis must also recognize that this decision-guiding procedure can be misused and perverted. For example, many critics in the United States have protested the use of unrealistically low discount rates in cost benefit computations as well as the dubious inclusion of uncertain benefits. Calculations based on these arbitrary variables have been used to justify the construction of otherwise untenable projects (Kneese and Smith 1966, pp. 291, 311).

The use of the cost benefit analysis involves other variables that are equally difficult to measure. For example, a Soviet economist, P. Oldak, has raised an intriguing question: How much allowance in the cost benefit analysis should be made for the pollution that a factory (once it is open) will inflict on itself (*LG* 6/3/70, p. 11)? Unlike the extra costs that are likely to be suffered by others because of a new plant's pollution, the extra operating costs that will be incurred by the new plant from its own pollution are usually ignored in studies of the costs and benefits of that prospective plant. As an illustration of this, Oldak questions whether or not the

designers of the pulp and paper plants at Lake Baikal
made adequate provision for the fact that the cost of
pure water to the plants would rise once they started
operation. The original rationale for the use of sites on
Lake Baikal was that the cost of operation of the plants
would be very low because the water necessary for pulp
and paper production would be so pure. The original
plan, however, provided for the dumping of the treated
waste water from the plants back into the lake. As we
shall see in Chapter 6, subsequent studies of the lake's
circulation pattern show that the waste water from at
least one of the plants is flowing back in a circle to the
plant's water intake pipe. The planners either did not
originally study the lake's circulation patterns, or they
felt it unnecessary to include such considerations in their
calculations. In any case, the quality of the water in the
vicinity of the plant has deteriorated, thereby necessitat-
ing supplementary treatment and expense and reducing
the economic feasibility of the plant (*Prav* 2/16/69, p. 1).
Undoubtedly, this type of "masochistic pollution" is
fairly widespread throughout the world. It is probably
equally certain that no allowance is made for such costs
when evaluating a project's future performance.

What enters into a particular cost benefit analysis is
often the personal judgment of the particular decision
maker. As a result, a cost benefit calculation can be used
to prove almost anything, often to the chagrin of conservationists who sometimes look to this device as a
means of supporting their cause. The Russians are as

vulnerable to such manipulations by their engineers with their cost benefit ratios as we are to our Army Corps of Engineers with their justifications.

Difficulties Peculiar to the USSR While most of the shortcomings of the cost benefit analysis we have been discussing so far apply to any user, there are some hazards that are unique to the Soviet economy. To be successful, the cost benefit technique also assumes that the economy's prices provide some kind of meaningful measure of scarcity relationships between commodities. While prices more often than not reflect such scarcity relationships in the United States, this is rarely the case in the USSR. The price reforms of July 1967 helped a little, but authorities in the State Committee on Prices acknowledge a continuing disparity. In fact, they sometimes even deny the desirability of such a pricing arrangement for the USSR (*EG* 2/68, No. 6, p. 10).

Examples of Cost-Benefit Misuse in the USSR
With all the shortcomings of the cost benefit type of analysis, it should not come as a surprise to learn that when such analysis has been used in the USSR, the findings have sometimes been used to support grandiose schemes of questionable merit. Thus as part of his campaign to defend the diversion of the Amu Darya and Syr Darya rivers for irrigation, V. L. Shul'ts, as explained in more detail in Chapters 7 and 8, has used the cost benefit analysis to justify the gradual disappearance of the Aral Sea. According to Shul'ts, the diversion of the rivers would make possible the irrigation of 8 million hectares (20 million acres) of new land. Crops on this land could

yield a profit of 750 million rubles, while the disappearance of the sea would merely result in a loss of about 60 million rubles in fish and about 10 million rubles from muskrat skins (*Kom Prav* 9/16/65, p. 2; *Kom Prav* 11/22/68, p. 2). On the face of it, the USSR would be better off economically if the water were to be used "productively" for irrigation rather than "wastefully" in a low-productivity sea for fishing.

Naturally the Aral Sea fisherman have been joined by many others to protest the implications of such calculations, but they are all hard put to crank their instincts into any kind of economic formulation. As many before have found, it is hard to place a monetary value on the recreational value of the Aral Sea, though it obviously counts for something. Another argument that may be somewhat novel for most Westerners is that the disappearance of this sea would probably set off a major alteration in the existing climatic pattern. Several geographers have warned that the elimination of this body of water would increase the continentality of the climate, which would mean longer winters and possibly shorter, hotter summers. While it might be possible to calculate how much this might reduce the value of the agricultural harvest and how much higher heating and clothing costs would be, it would be considerably more difficult to attach a ruble value to the psychological and physiological effects of such a transformation.

Even when a cost benefit calculation indicates that a particular project is economically unjustified, the planner can always juggle the figures around until he obtains the necessary answer. This is particularly tempting when prices and land values, as in the USSR, are frequently set arbitrarily and are often not related to scarcity and market forces. Thus many economists have charged that existing yields on land in the USSR are often intentionally understated to make land confiscation, flooding, and dam construction look more justifiable (*Prav* 3/10/70, p. 2). Similarly, to justify the construction of the pulp and paper plants on Lake Baikal, the planners in the Ministry of Paper and Timber Industries intentionally understated the costs of production in order to demonstrate the future profitability of the plant. It was ultimately acknowledged that the plant would cost at least 22 million rubles more than originally admitted (*LG* 4/15/65, p. 2). In sum, for the conservationist, be he Russian or American, the cost benefit analysis can be a valuable tool for eliminating specious proposals, but there is always the danger that the cost benefit technique may be distorted into more of a dangerous weapon than a protective restraint (Streeten 1971, p. 2).

The "Department of Public Works Mentality"

The self-serving activities and virtually unrestrained power of state agencies we have been discussing represents what can be called "Department of Public Works-ism." It is usually reflected in the feeling "Build We Must," not because of the need, but because we want to show that we are earning our salary and can erect monuments to our energies. Like the railroad bridge over the river Kwai, the project must be completed or expanded irrespective of the end purpose. The sense of satisfaction comes as much from the challenge and process of construction itself and the manipulation of all those men and materials as it does from the final project. Again this attitude exists in both capitalist and socialist societies. However, it is more of a threat in the USSR because almost all construction is generated by the state, usually without private cost benefit analysis, not to mention a social cost benefit calculation.

The havoc that can be wrought by this type of approach is illustrated by what happened at Lake Balkhash, the largest lake in the republic of Kazakhstan. The first plans for redoing the water regime of the area apparently were developed in the early 1960s. In a report submitted to officials in Moscow in March 1965, it was formally proposed that a large dam be erected at Kapchagaisk on the river Ili. The Ili is the major river in the area and the source of 75 percent of the lake's water. It was asserted that the new reservoir would make it possible to irrigate 430,000 hectares (about 1 million acres) of land, 250,000 (about 600,000 acres) of which would be devoted to rice. The newly irrigated land would provide a yearly revenue of 240 million rubles. Fish would be bred behind the Kapchagaisk Dam, and they would yield another 2.98 million rubles. Water transportation would be facilitated so that shipping costs would be lowered by as much as 90 percent. In all, the heightened revenue would make possible a payback period of 1.5 years.

Shortly after the project was announced, doubts about the wisdom of the proposal were heard from the residents of the town of Balkhash. Situated midway on the northern shore of the lake, Balkhash was originally a desert, but residents of the town turned it into a flourishing center with several large factories and a population of 77,000. The lake serves the residents of the city as a source of drinking and industrial water, a breeding place for fish (150,000–160,000 centners a year, or about 16,500–17,600 tons) and as a recreational area (*LG* 12/17/69, p. 11). Anything that affects the quality of the lake threatens the very existence of the town. Unfortunately, Lake Balkhash is very vulnerable. Because of its arid location, water in the lake evaporates rapidly. In the western half of the lake, which just includes the town of Balkhash, there is no serious problem as long as the Ili provides a constant flow of fresh water. At the eastern end of the lake, however, the salt content of the lake is very high. As work on the dam moved ahead, the residents of the area began to realize that the flow of the Ili would be seriously disrupted when it was blocked to fill the Kapchagaisk reservoir. The diversion would mean a reduced flow of fresh water into the lake. This in turn would lead to an increase in the lake's salt content and the poisoning of the town's fresh water supply.

As protests began to mount over the Kapchagaisk Dam, it was discovered that the Ministry of Power as well as some engineers in the republic of Kazakhstan had engaged in all kinds of subterfuge to justify the economic value of their project. Gosplan in Kazakhstan had questioned the soundness of the whole plan as early as November 16, 1964, six months before the initial design

had been submitted to Moscow. It soon became apparent
that the potential return from agricultural development
had been overestimated and that the yield would be
much smaller. Moreover, it was also acknowledged that
if the dam held back the spring floods, there would no
longer be any flooding of the delta area. This meant that
over 300,000 and maybe as much as 600,000 hectares
(750,000 to 1,500,000 acres) of presently productive land
would turn into a desert in two years (*LG* 2/11/70, p. 11).
Recognizing that the agricultural benefits of the project
could no longer be stressed, the Ministry of Power and
the construction engineers shifted their emphasis to the
benefits that would accrue from the creation of a recreational area nearer to the city of Alma Ata and the electric
power from the dam. Neither the recreational value nor
the power potential was even mentioned in the 1965
plan. Furthermore, once construction had been approved,
another calculation was made, and it was discovered that
it would take four years, not one and one-half years, for
the dam to pay for itself. At the same time, the height of
the dam was raised. This increased not only the storage
capacity of the reservoir but also the amount of water
that would have to be diverted from Lake Balkhash.
Some critics pointed out that only 6.6 cubic kilometers of
water would have to be used to generate the electricity
needed for the surrounding area; nevertheless, the
capacity of the dam was designed to hold 28.1 cubic
kilometers. With the higher dam, the capacity of the
power station grew from 250,000 to 360,000 to 434,000
kilowatt-hours. One critic points out that the local

Gosplan had argued for a lower level, but the height of the dam was increased anyway, because the construction planners did not want to risk the criticism that the generating capacity of the power station would be too low to justify its construction. "The planners did not want to worsen their indicators." (*LG* 2/18/70, p. 13).

While debate over the size and impact of the dam continued, so did construction. The dam was completed in September 1969, and it began to store the water from the spring floods in 1970. After an emergency commission had been appointed, it was finally decided to restrain the dam builders somewhat. Instead of trying to fill the reservoir in five years, the hydroelectric authorities agreed to wait eight to ten years. This means that the flow into Lake Balkhash will not be reduced as much as initially expected. Yet it will still be a larger diversion of water than other specialists had recommended. They had proposed that the reservoir be filled in thirteen years (*LG* 3/25/70, p. 12; *LG* 7/11/70, p. 2). In addition to taking a longer time to fill the dam, it was also agreed to reduce the height of the dam from 489 to 485 meters. This will also reduce the amount of water that will be held back. While this is an improvement, it still does not meet the demands of some critics who wanted the level lowered to 475 meters. Those who have argued for lower levels were apparently correct. Confirming the fears of the pessimists, the average mineral content of the lake's water near the town of Balkhash in the first half of 1970 rose by 8 percent over the preceding year. Moreover, despite the formal agreement to wait eight to ten years to fill the reservoir,

Russian conservationists were dismayed to learn that by the end of the first year, the reservoir was already one-quarter full (*Kaz Prav* 9/10/71, p. 2).

The events at Lake Balkhash fill one with a sense of *déjà vu*. The engineers and contractors have a project. Criticism only spurs them on to greater efforts. There appears to be some speculation that the dam on the Ili River was put forward because it was the only major river in Central Asia that did not already have a major dam. As will be seen in Chapters 7 and 8, the other two rivers, the Syr Darya and the Amu Darya, have already been dammed, shunted, and exhausted so that there is little left of them to work with. The fact that initially there was no need for electricity from the Kapchagaisk Dam and that other sources of electricity for Alma Ata were available at a cheaper cost was of no consequence. As one engineer put it, "We were assigned the job of completing this work [the Ili River Dam] and we have fulfilled it!" (*LG* 3/25/70, p. 12.)

Nor is Lake Balkhash unique. At Yasnaya Polyana, Tolstoy's old country home, a similar penchant for construction has led to the near destruction of a historic and treasured forest. In 1955 a coal gasification plant called Shchekino was built about a mile from the estate. As part of Khrushchev's campaign to develop and expand the chemical industry in the early 1960s, a urea plant was added to the Shchekino complex. As soon as the plans were announced, the people at Yasnaya Polyana began to protest that the fumes from the factory would harm the vegetation on the estate. As Iu. Fedorova, the chief

agronomist of the estate put it, "Just as the conservationists at Lake Baikal are now being reassured, so we
agronomists were told that the chemical factory would
cause no damage at Yasnaya Polyana." (*LG* 3/15/65,
p. 27.) Shortly after the urea plant went into operation,
however, the people at Yasnaya Polyana noticed that
their famous pine forest was dying. After more protests,
a special commission was established by the Council of
Ministers of the Russian Republic. Numerous remedies
were suggested. But as Fedorova later explained, "The
special commission has closed down, but now it appears
that already new work has begun—the work of the
chemical combine." Summing up the whole process, she
lamented,

The tactics of predators are always the same. First they
promise to save, guard, preserve in full splendor and
even expand the threatened object. But after the damage
occurs, then they plead the importance of the higher
governmental priorities. Now the engineers rationalize
their damage with statements like "Think of it—3 pine
trees have been killed. What is more important for our
government—3 pine trees or a kapron [synthetic fiber]
plant?" (*LG* 3/15/65, p. 2.)

Gross National Product versus "Net National Well Being"

Related to Department of Public Worksism, yet distinct
from it, is the drive for faster economic growth with its
resulting effect on nature. Department of Public Worksism
exists both in an environment where economic growth is
unimportant and in one where it is the prime considera-

tion. However, in a country like the USSR, where it is generally agreed that economic growth must be accelerated in order to overtake other more advanced countries, the attack on nature is usually pursued with much more intensity.

In underdeveloped countries, raw materials are often the only products that attract foreign exchange. The dollars and hard currencies dangled in front of developing countries are usually tempting enough to win over those few who might be worried about their grandchildren's natural endowment. Pre- and postrevolutionary Russians were no more immune to these temptations and necessities than anyone else. In fact, because of the immense size of the country and its rich endowment, there seemed to be plenty of resources available for exploitation and plenty of room to discharge those wastes that were generated in the process.

One of the major economic "contributions" of the USSR has been to make the rest of the world growth conscious. Until the start of Russia's five-year plans, most of the world's inhabitants seldom concerned themselves much with national economic growth. Everyone concentrated on his own personal or corporate growth, and somehow the macro picture did not stimulate much interest. The Russians changed that. Moreover, in order to stimulate as rapid growth as possible in an underdeveloped environment, the Russians designed an incentive system that placed maximum emphasis on increasing production. Everything else was secondary. This was reflected in practice as well as in spirit.

It usually did not matter what it was that was being produced as long as there was more of it this month, quarter, or year than last. Premiums were dependent on fulfilling 100 percent of the target, and bonuses were increased progressively as the plan was overfulfilled. Cost control was of minor concern to the manager since his premiums, at least until 1966, were practically independent of how high costs were. In such a climate, it seemed only natural that the country's resources would be used wantonly.

There are numerous illustrations of the impact of such a policy on the environment. Major natural treasures as well as daily work practices have been affected. Among the major projects, for example, was the transformation of Lake Sevan from a natural wonder into part of a hydroelectric scheme. One of the earliest activities undertaken in Armenia after the revolution was the construction of a major hydroelectric station that used the water of Lake Sevan, one of the highest lakes in the world. The initial justification for taking the water from this beautiful and unique lake and carrying it through a tunnel to an underground power station was that the need for electricity in Armenia was of critical importance and the keystone to the whole policy of industrialization. Conservationists in the area were very much opposed to the whole idea for fear that the tunnel would draw off so much water that the level of the lake would fall sharply. Because of such protests, conservationists were able to block construction of the project for several years after plans for it were first drawn up before the revolution

(Maryan 1968, p. 40). After the revolution, however, the emphasis on economic growth and the need for electric power proved to be too strong for the efforts of the conservationists.

With time, the lake's waters were also seized upon for use in irrigation. The effect of all this water loss brought about a 17-meter (56-foot) drop in the level of the lake and a loss of 38 percent of its volume and 13 percent of its area from the prerevolutionary norm (Gambarian 1968, p. 89). This set off all kinds of effects including the serious erosion of the lake's shore. In addition, the chemical and biological makeup of the lake has changed. There has been an increase in the quantity of organic matter and a decrease in the pH levels. This, in turn, has accelerated the process of eutrophication.

To remedy the situation, water from other areas is being diverted at great expense to Sevan, and, increasingly, natural gas instead of water is being used as a source of electric power. As a result, the level of the lake is no longer falling as rapidly, and in fact in 1969 it even rose 47 centimeters (about 20 inches) (*Prav* 8/23/69, p. 6). But all of this has taken its toll on nature, and the cost of the remedy will be at least $100 million.

In the chapters ahead when I describe what has happened at places like Kislovodsk, the Black Sea coast and Lake Baikal, I shall cite numerous other examples of where the environment has been similarly sacrificed for the sake of economic growth.

The attitude that nature is there to be exploited by man is the very essence of the Soviet production ethic not only in major projects affecting nature but in the day-to-day work attitudes in the factory. In moralistic tones reminiscent of the nineteenth-century Presbyterians, the Soviet workman is urged to increase production and manifest his destiny over nature. Inevitably this clashes with the conservationists' ethic—not to mention the environment. Several Russian economists have lamented the practice of Soviet managers who have willingly incurred whatever fines may be imposed for pollution in order to fulfill the plan and thereby earn a premium. "The manager would rather pay a fine of 500 rubles [$555] if it means he can earn a premium of 5,000–10,000 rubles for overfulfilling his plan." (*Sot In* 8/15/70, p. 2.)

With the overriding pressure to increase production, there is usually little incentive or encouragement to spend any money on pollution-control equipment. Money spent on pollution control is usually a nonproductive expenditure. It seldom generates either increased production or increased profit for the factory. In fact, money expended on pollution control reduces the amount of money available for machinery to increase production. The factory manager thus is extremely reluctant to divert any of his investment funds away from "production" toward "conservation." Because of the existing incentive system in the USSR, his bonus would probably suffer if he followed the conservation path. The choice is well summed up by one plant director who was upset because complaints had been made about his plant's pollution. Demands had been made that he install treatment equipment. "It is always simpler to demand. But what about the plan? Are you going to order the plant to cease production? That is the dialectic. One has to choose between civilization and one's love of nature." (*Iz* 6/27/70, p. 4.)

The reluctance of Russian factory managers to request or use investment funds for pollution control is a perplexing dilemma for Soviet conservationists. In instance after instance, complaints are published that Soviet industry has failed to fulfill the planned installation program (*Prav* 8/23/70, p. 3; Belichenko 1968, p. 112). Seldom are such targets more than 75 percent fulfilled in the RSFSR (the Russian Republic), and in 1963, in Turkmenistan, it was fulfilled by only 6 percent (*Iz* 2/4/67, p. 3; *VST* 4/70, p. 11; Ivanchenko 1969, p. 240). It is much more important to the factory manager that he see that new plant production start on schedule. Pollution-control operations, which logically should be ready at the same time, invariably are accorded much lower priority and so lag far behind production operations. This happened at Lake Baikal as well as numerous other places. At Orenburg, when the purifiers for a steam-generating plant were delivered one and one-half years after the plant was put into operation, there were no complaints from the local government authorities about the lag. In fact, the plant and its managers were rewarded with premiums for an early opening of plant operations (*Kom Prav* 7/24/70, p. 2; *Sot In* 1/28/70, p. 1). This attitude is reflected in a discussion with a plant official reported by the writer Vladimir Soloukhin in his travelogue, *A Walk in Rural Russia*.

If the factory did not carry out its obligation according to plan, there would be recriminations—someone might even lose his job. But if the filter were not made, no great harm would be done, no inquiries would be made, no one would notice. The fish would disappear? People would fall ill because of the water? Well, in the first place, it is never certain what is the cause of people falling ill. Our job is production according to the plan. (Soloukhin 1967, p. 86.)

This same attitude holds at the national level since, until recently at least, the national targets have also been set in terms of physical quantities of production. Consequently, when an occasional factory manager decides he wants to spend money on pollution control, he runs a risk. This was the experience of the director of the Makeyevka factory in the Donbass. He insisted on an allocation of funds for a water-treatment plant. The Ukranian Ministry of Ferrous Metallurgy reprimanded him. He was accused of having an incorrect attitude toward state funds. By the time the equipment was finally authorized and installed, the factory had already polluted the nearby streams (*CDSP* 5/25/60, p. 21).

Instead of serving as a referee between polluters and conservationists, government officials usually support the polluters. It is necessary to remember that the state *is* the manufacturer, and so there is almost always an identity of interests between the factory manager and the local government official. The most important criterion for any government official who seeks promotion or recognitition is how much his production has increased in his region, *not* to what extent his rivers have been cleaned up this year. It follows that few government officials are likely to be particularly sympathetic to those who threaten the attainment of new production records. As Zile puts it, this is one of the hazards of being both the police and the policed (Zile 1970a, p. 9).

Even if the main targets are broadened to include such control devices as water aerators and electrostatic precipitators, it is still unlikely that there will be a rush to install them as long as such devices do not increase the production yields of the prospective purchaser. Installation would probably be dependent on either strictly enforced legal requirements or the substitution of some new measure such as "Net National Well Being" (NNWB) for the traditional Gross National Product (GNP) and physical production targets. With a concept like NNWB, the produced value of such items as steel and chemicals would be included only after a deduction has been made for the cost of the pollution that has been generated in the course of their production. Economists around the world who have been calling for such a concept are now being joined by a few Soviet economists, such as P. Oldak. He is even prepared to see a 7 to 10 percent drop in the amount by which the economy grows each year (*LG* 6/3/70, p. 11; Tsuru 1970, pp. 54–57; Tsuru 1971, p. 5).

The Vested Interest of the Private Property Holder
In any evaluation of the factors determining environmental disruption in communist and noncommunist societies, one of the more intriguing questions concerns

the role of private property. The absence of private land speculators and manufacturers in a communist economy presumably means that there will not be the havoc and environmental disruption such groups have created in a capitalist society. Private ownership of property often does create environmental disruption. If the private property owner calculates that the offer he receives from the oil drilling company for his land will provide him with a larger benefit than he presently derives from his property, he will sell, and the land will be converted to industrial purposes. This usually depends on how large the potential private benefits for the property holder will be. He takes little or no recognition of the benefit society as a whole has been deriving from this property. But by the same token, the interests of the private property holder may occasionally prevent the transfer of land to industry. If the private owner receives an offer to sell his land, he may keep it in its natural state if the price is not high enough. For example, if the prospective buyer can use the land only for a quarry or a restaurant, he would probably be forced to offer less than if he expected to find oil. In that case, the private property holder may find that the price offered him does not equal the private benefit he presently derives from the property in its existing state. Thus, on some occasions, he may retain the land in its existing state.

No such restraint exists in the USSR. Conceivably even without private property the government could exercise its authority and assign ownership or guardian authority over a specific site, and theoretically this should help preserve it. In some cases this does help, especially if the state protector stands to suffer economically or if there is some valuable collection of flora and fauna that seems unique. But usually where there is an economically valuable raw material available, exploitation of the raw material is so tempting and the interest of the guardian agency so modest that the protection is not very effective. Consequently there are many instances in the USSR where the economic value of the land in its proposed alternative use is actually lower than the value that it generates in its existing state. In sum, private landowners do often decide to sell their land for nonconservation purposes. Then the private benefits from the sale by the former owners are high, and the social costs are ignored, as always. The Russians, however, under their existing system now have to rely almost solely on benefit calculation to prevent environmental disruption; they lack the first line of protection that would come from balancing private costs and private benefits.

As we shall see in Chapter 5, all of this helps to explain why state contractors in the Georgian Republic continue to haul vast quantities of sand and pebbles away from the Black Sea beaches for use in building construction. Resolutions are passed year after year prohibiting such activities, but the practice persists because the contractors regard the beach as a free good. Since there is no one with private property rights (a resort owner or a private homeowner) to say that the beach has a higher value as a beach, no one has been able to put a stop to the practice of hauling the beach away piece by piece. As a result, the seashore has been further eroded as the sand and pebbles that cushioned the powerful pounding of the waves have been used to build resort hotels and a sea wall on what once was a world-famous beach.

Similarly, Soviet economists and geographers have been complaining with increasing frequency that vast areas of land are being flooded in order to build hydroelectric plants. Some of the critics argue that these new power plants are often less of a contribution to the country's economy than the now vanished land (M. I. L'vovich 1969a, p. 110). The same thing happens, of course, in the United States, but the prices are more likely to reflect existing scarcity relationships. Also the state in the United States generally exercises power over only a small portion of the economy. Instead of the state gathering up the land, it is usually the private corporation that is the moving force. In any case, both the state and the private corporation normally have to deal with other private property holders who often disagree with the new plans and are prepared to fight them. In the USSR, however, the state is the power behind all such changes, and therefore it is limited by few restraints. A massive reordering of nature is more likely to take place in a society where all the land and means of production are owned by the state and there is no private ownership to suffer losses, create restraint, and voice objections.

Private property sometimes also provides another check to environmental disruption. One impediment to any effort to assign accountability for social costs in either a capitalist or socialist society is that the damages are often spread so widely that no one feels aggrieved enough to initiate a complaint or action. In other words, the damages may be so diffused that no one party suffers enough economic damage to make it worth his while to take preventive action. The likelihood of a protest or a court case is stronger, however, if one particular party, preferably an influential one (that is, another manufacturer), ends up bearing the brunt of the damage. In a sense, we can distinguish two subcategories of social cost: private social costs—those that are primarily borne by one or a few sufferers—and public social costs—those that are spread out over a wide area so that no one is made to suffer inordinately in either physical or economic terms.

In a socialist society it would seem that it would be more difficult to stimulate preventive action in both the case of public and private social costs. Because private land ownership is prohibited in the USSR, the individual has less of a vested interest in fighting the construction of a new factory in his neighborhood or the mining of some raw material in the area. Except when a state-owned factory finds that its operating costs are substantially and directly affected by another factory's pollution, protest must depend on social consciousness, and not on the actions of private property holders who respond out of the fear of a private loss. Of course, social consciousness can be very effective, as has been demonstrated by the success of such groups as the Sierra Club and the League of Women Voters. Nevertheless, the elimination of the private property holder and his accumulating instincts often means the elimination of the first line of defense against the expansion of environmental disruption. Sometimes this can be a very effective force.

Conclusion

Some of the factors considered in this chapter apply to all economies, some are particularly relevant to developing economies, and some to economies that are socialist or those in which all the means of production are owned by the state. It would be too much to say that the solution to environmental disruption in socialist societies such as the USSR is necessarily more difficult than in other societies. Nonetheless, many of the theoretical advantages that a socialist society would seem to have for coping with the problem have proved to be illusory in practice. Moreover, some of the existing advantages are offset by economic forces that actually tend to generate environmental disruption. Based on the Soviet experience so far, there is no evidence to indicate that nationalization of all the means of production in one country or throughout the world will provide any cure all for environmental disruption.

3 Water Pollution

In the previous two chapters we discussed the ideological underpinnings of environmental law in the USSR as well as the economic and political realities. We now consider various aspects of the Soviet environment in greater depth.

Water Availability

Because it is so large, the Soviet Union has enormous quantities of fresh water at its disposal. Its rivers discharge as much as 470 cubic kilometers of water a year, which is about one and a half times the amount discharged in the United States (*Prav* 1/5/70, p. 2; *Prav* 5/31/71, p. 3). Such estimates, however, can hardly be regarded as precise. Even the estimates of the Soviet geographer M. I. L'vovich, who has specialized in calculating water balances for the world as well as the USSR, have been subject to change. In the early 1960s he concluded that the total river flow was 4,220 cubic kilometers of water (M. I. L'vovich et al. 1963, p. 48). By 1969 he had revised all his figures upward so that, according to his data in Table 3.1, total precipitation for the USSR amounted to 10,960 cubic kilometers, and total runoff was 4,350 cubic kilometers (M. I. L'vovich 1969a, p. 101). Of this, about 1,000–1,020 cubic kilometers appears on the surface.

With 10 percent of the world's precipitation and with more than 10 percent of the world's rivers flow moving through the USSR, presumably the Russians should be well enough supplied with water to quench their varied needs. But the adequacy of this endowment is a bit deceptive. First, the Soviet Union occupies about 16 percent of the world's land mass, although it has only 10 percent of its precipitation (Shabad 1951, p. 3). Second, the bulk of the water available is located in the eastern part of the country, but the population is clustered in the western part. In contrast to the 155,000 rivers in Siberia, there are only 45,000 in the European part of the USSR (Vendrov and Kalinin 1960, p. 35). These Siberian and northern rivers carry about 62 percent of all surface water in the USSR into the Arctic Ocean and about 20 percent into the Pacific (*EG* 8/18/65, No. 33, p. 15). Together this accounts for about 3,240 cubic kilometers of the total flow (*Iz* 1/20/68, p. 3). This leaves the remaining 18 percent of the water flow to be shared by 80 percent of the population (*Iz* 12/12/71, p. 4). After allowing for the 4 percent of the total that goes to the Baltic, there is only 14 percent left to flow through the heavily populated and industrialized areas of the Ukraine, the Crimea, Donets Basin, the Volga Basin, Kazakhstan, and Central Asia to the Black, Azov,

Table 3.1 Estimate of Annual Water Balances over Land Surfaces of the World and the Soviet Union (in cubic kilometers)

	Total World Land	USSR
1. Precipitation	108,400	10,960
2. Total runoff	37,300	4,350
3. Underground (stable) runoff	12,000	1,020
4. Surface runoff	25,300	3,330
5. Effective moisture supply	83,100	7,630
6. Evaporation	71,100	6,610

Source: M. I. L'vovich 1969a, p. 101.
Note: Row 1 = row 4 + row 5; row 2 = row 3 + row 4; row 5 = row 3 + row 6.

Caspian, and Aral seas (*Iz* 1/20/68, p. 3; *Iz* 12/12/71,
p. 4; Armand 1966, p. 60). Although the average availability of water for the country as a whole equals 20,000
cubic meters per capita, in the Lena Basin in Siberia
there is as much as 500,000 cubic meters of water per
person. At the other extreme, in the Donets and Dneprovsk Basin, there is only 400–600 cubic meters per person. In Moldavia and the nondesert areas of Central
Asia there is as little as 300 cubic meters per capita (*EG*
8/18/65, No. 33, p. 15; M. I. L'vovich 1969a, p. 103;
Prav 5/13/71, p. 3). Finally, not only is the water poorly
distributed geographically, but it is unevenly distributed
seasonally as well. Sixty percent of the rainfall comes in
the spring and summer, so that much of the water
washes off as floodwater. Consequently, for much of the
year, many rivers are dried up (M. I. L'vovich 1962,
p. 4).

Use of Water Resources

Though 4,350 cubic kilometers of water is theoretically
available for use in the Soviet Union, about 170–200
cubic kilometers are used by man (Oziranskii 1968, p.
67). About 130 cubic kilometers of that total is used by
industry and agriculture. In the industrialized regions,
industry may consume as much as 70 to 80 percent of the
fresh water, especially for thermal cooling (*Prav* 1/5/70,
p. 2). The economist S. Oziranskii predicts that by 1975
the demand for water may reach 340 cubic kilometers a
year and by 1985, 580 kilometers, an increase of two and
a half to three times over that of 1965 and about ten

times that of 1940 (Oziranskii 1968, p. 67; *Prav* 5/13/71, p. 3). Others estimate that there will be a fourfold increase by the year 2000 (Mazanova 1969, p. 16; Abramov 1970, p. 6). As shown in Table 3.2, Oziranskii anticipates that from 1965 to 1985 there will be almost a fourfold increase in the use of water by the population, a threefold increase by industry and agriculture, and a fivefold increase for thermal use. In individual republics the need for future water supplies may be particularly acute. One authority estimates that by 1985 Moldavia will need fifteen times more water than it is using now (*Sov Mol* 6/11/69, p. 2). There are some who believe that the RSFSR will need five times the water it consumed in the late 1960s (Sushkov 1969, p. 4).

Table 3.2 Water Consumption (cubic kilometers per year)

	1965		1975		1985	
	Total	Not Returned to System	Total	Not Returned to System	Total	Not Returned to System
Urban populations and suburbs	8.4	3.8	16.2	6.1	33.5	9.5
Industry	22.2	6.9	38.1	12.6	66.0	27.5
Heat and energy	18.4	1.5	39.5	2.0	101.9	4.7
Agriculture	6.8	6.5	10.0	8.7	17.1	13.3
Irrigation	114.8	92.2	168.1	144.1	255.8	214.1
Fishing	16.3	4.2	45.6	12.0	65.7	17.3

Source: Oziranskii 1968, p. 67.

Because of the recycling of water the burden is not as
heavy to bear as it seems at first. Thus out of the 102
cubic kilometers of water that it is anticipated will be
needed for the generation of electricity, only about 5
cubic kilometers will actually be exhausted. Overall it is
estimated that Soviet industry recycles 57 percent of all
the fresh water it uses (Shabalin 1970, p. 16). Only
when water is used for irrigation is a large portion of it,
about 80 percent, exhausted so that it cannot be reused
(Oziranskii 1968, p. 67).

Needless to say, the USSR is not unique in anticipating
a greater need for water in the years to come. Specialists
in the United States have estimated that demand for
water use in the nation will have doubled from about
320 billion gallons per day in 1960 to 600 billion gallons
per day in 1980 (Weinberger et al. 1966, p. 139). That
is somewhat less than the growth expected in the USSR.
Per capita consumption in the United States is presently
higher than it is in the USSR; thus the Russians presumably have further to go as they continue their
industrialization.

Whether it be in the United States or the Soviet Union,
increased consumption of water almost inevitably means
that there will be an increase in sewage and the amount
of water to be treated. Understandably, it is hard to
ascertain estimates of what these figures are and will be.
Water discharge is even less well monitored than water
intake. Consequently, as in America, Soviet estimates of
present and potential sewage discharge vary widely.
Presumably, the most reliable study is the one that has

been prepared by a commission of the Soviet Academy of Science. It found that, in 1969, 99 million cubic meters (why not 100 million cubic meters?) of unclean or inadequately treated water was discharged each day into the country's rivers and other water bodies (Buianovskaia 1969, p. 77). This is equivalent to 36 cubic kilometers a year.[1]

If the amount of sewage discharged continues to grow at the present rate, as is expected, there may be a doubling in volume every ten years (A. I. L'vovich 1963, p. 35; *Kom Tad* 3/7/71, p. 3; *Prav Uk* 8/29/67, p. 2). Although again the data are fragmentary, one authority estimates that the discharge of liquid wastes will total 60 cubic kilometers in 1980 (*SGRT* 1968, No. 9, p. 769). Some see a growth of 2.5-fold from 1969 to 1980, while others expect sewage discharge to multiply five times by 1985 to 1990 and perhaps fifteen times by 2000 (M. I. L'vovich 1969a, p. 114; *Prav Uk* 1/11/69, p. 2). To the extent that the estimates of water consumption and sewage discharge are comparable, the discharge of sewage will grow at a considerably faster rate than water consumption. This is a normal result of the growing complexity of industrial technology and production. Similar findings have been made regarding the United States (Weinberger et al. 1966, p. 133; Commoner et al. 1971, p. 3). In both countries, there is concern about when and by how much demands for water consumption

---

[1] Other estimates for about the same period ranged from 18 cubic kilometers to 27 cubic kilometers a year (Ivanchenko 1969, p. 239; *Iz* 6/27/70, p. 4).

will exceed available supply (Weinberger et al. 1966,
p. 138; A. I. L'vovich 1963b, p. 35; M. I. L'vovich
1969a, p. 114; EC 1/67, No. 4, p. 37; SGRT 1969, No. 9,
p. 769). This, in turn, is partially dependent on the
amount and degree of treatment to which the sewage is
subjected. As we shall see, the Russians have much further
to go in providing treatment than we do in the United
States (Ivanchenko 1969, p. 240).

Water and Sewage in Moscow

An appreciation of the difficulties confronting Soviet
sanitation authorities can be gained by a study of how
Moscow has handled its water supply and sewage needs
over the years. Water treatment procedures in other
major cities and republics of the USSR will help to
provide additional insight.

Prerevolutionary Development Apparently the first
effort to provide a centralized water supply for Moscow
began with the construction of the Mytushchinskii water
project in 1804 (Gorin 1968, p. 35; Galanin 1967, p. 34;
D'iachkov 1967, p. 30). The initial system was refitted
and expanded to provide 6,000 cubic meters of water a
day in Moscow from 1853 to 1858 (Abramov 1970, p. 5).
There is less information available about the origins of
the centralized water system in St. Petersburg, but its
water system was also reconstructed a few years later, in
1863–1864, and its first pumping station was opened at
the same time (Abramov 1970, p. 5). The capacity of the
Moscow system was expanded to 18,000 cubic meters a
day in 1892 and 43,000 cubic meters in 1902–1903

(Gorin 1968, p. 35; D'iachkov 1967, p. 30). A modern water supply system, however, began only with the establishment of the Rubleva water works on the Moscow River in 1903. This plant was designed to service 4,200 homes. It fed into a pipe network 390 kilometers long (D'iachkov 1967, p. 30). By the time of the revolution, the system as a whole was providing 170,000 cubic meters of water a day or about 80 liters per person to 9,163 homes through a pipeline 692 kilometers in length (Gorin 1968, p. 37). However, this system supplied only the residents of the center of the city; everyone else relied primarily on 140 wells scattered throughout the area. During the winter half of these wells froze (D'iachkov 1967, p. 30).

Almost a century passed between the construction of Moscow's first centralized water system and the installation of its first sewer. It was not until 1898 that a sewer was made available to the central part of Moscow. The sewer served to replace the sanitary brigade, which even then continued to collect night soil in the outer parts of the city (D'iachkov, 1967 p. 34). The first plans for the sewer were proposed in 1873, and the first portions of the sewer were put into operation in 1893. However, it took five more years before the whole system could be put into operation. Initially the sewer served only 219 households. By 1905, 6,000 households were included (Matvev 1969, p. 2). By 1917, the system stretched a distance of 534 kilometers and had a capacity of 90,000 to 108,000 cubic meters a day (Galanin 1967, pp. 34, 36). Not surprisingly, no treatment plant was built to process the sewage. In a continuation of the sanitary brigade process, much of the sewage ended up in sewage farms rather than being dumped directly into the water. The Liublinskie sewage farm took 49,000 cubic meters a day, and the Liuberetskie sewage farm took 41,000 cubic meters (Galanin 1967, p. 34).

Water supply and sewage conditions throughout the rest of prerevolutionary Russia were not much better. Water was supplied centrally in about 215 Russian cities to certain selected areas, usually upper-income neighborhoods (*VST* 11/67, p. 3). For the most part, sanitary conditions were quite unsatisfactory (Matveev 1969, p. 1). Sewage facilities were even more primitive or limited in scope. Only about eighteen of the larger cities such as Moscow, Kiev, Odessa, and St. Petersburg had anything like a sewer system, and in only a few of them were there even provisions for carrying the waste to sewage farms (Zhukov 1970, p. 9). A group called the "Temporary Committee to Find Measures to Protect Water Reserves of the Moscow Industrial Region from Pollution of Sewage and Wastes from Factories and Plants" was set up in 1912 to foster professional concern for water sanitation (Zhukov and Mongait 1968, p. 13). Like the sanitation network itself, however, the activities of this group were ineffectual.

In the confusion of the revolution and the period of War Communism that followed, the sanitation capabilities within Moscow diminished. As shown in Table 3.3, major changes had to wait until the 1930s, and major improvements especially in sewage capacity were not completed until the 1960s.

Table 3.3 Moscow Water Supply, Sewers, and Sewage Treatment
Capacity

	Water			Sewage	
Year	Pipeline (kilometers)	Water Discharge (cubic meters/day)	Per Capita (liters/day)	Pipeline (kilometers)	Capacity (1,000 cubic meters/day)
1858	—	6	—	—	—
1893	—	12.5	—	—	—
1902	—	43	—	—	—
1903	390	44	—	—	—
1917	692	—	88	534	90–100
1920	—	150	—	—	—
1927	—	—	—	—	—
1930	914	287	110	590	160
1935	1,088	561	163	—	—
1937	—	—	—	797	282
1940	1,480	1,053	243	—	—
1945	1,543	1,153	347	—	—
1947	—	—	—	970	520
1950	1,649	1,444	321	—	—
1952	—	1,500	350	—	—
1955	2,066	1,979	400	—	—
1957	—	—	—	1,315	1,037
1958	—	—	—	—	—
1960	2,910	2,716	—	—	—
1963	—	3,200	500	—	—
1965	3,821	3,524	545	—	2,740
1966	4,000	—	—	—	—
1967	4,060	3,644	560	2,490	3,180
1970	5,000	—	—	3,000	3,145
1971	—	—	—	—	4,120 (plan)
1980	—	—	1,000	—	8,000 (plan)

Sources: Galanin 1967, p. 36; *VST* 11/67, p. 3; *VST* 2/69, p. 41; Gorin 1968, pp. 35–37; Ivanov 1970, p. 18; Chvanov 1971, p. 15; Drachev and Sinel'nikov 1968, p. 285; D'iachkov 1967, p. 30; *SN* 7/13/71, p. 207.

Postrevolutionary Development FRESH WATER The prerevolutionary pumping station at Rublevsk had to serve as the centerpiece of the water supply system for several years after the revolution. While study groups were created shortly after the revolution to coordinate and conserve existing water supplies, little was done to change or supplement the basic system until the 1930s. Much of this delay was due to the confusion that existed during the civil war and to the inability and unwillingness to mobilize investment on infrastructure when the country's main efforts were directed to building up industry. Another important factor complicating the expansion of freshwater supplies was Moscow's location. There may have been historical and nationalistic reasons for keeping the capital where it was, but from the point of view of water supply and disposal, Moscow is located on a poor site. The area of the city was never meant to support a large population. Its water resources are too limited. The main source of water for the city is the Moskva (Moscow) River, which is a relatively small stream (Drachev and Sinel'nikov, p. 284). In the immediate vicinity there is too little water to draw from and no place to put the sewage of 7 million people.

Most of the original settlers of today's largest cities never imagined that millions of people would someday follow in their footsteps. Whether a city will grow is not merely a matter of its water resources. Nevertheless, beyond a certain point water availability can be a brake on further development or at least a very expensive consideration.

While Berlin, Paris, Dallas, and New Delhi are other inland cities that have grown to become major metropolises, Moscow is the largest, the most heavily industrialized, and the furthest from major water bodies.

Despite Stalin's unwillingness to divert large sums to the niceties and even the necessities of life, he was forced to act by the inadequacy of Moscow's water supply. In 1926 there were already 2 million Muscovites, and only a little more than half of them drew their water through pipes at home. About half a million had access to "good" pumps or hydrants, but as many as a quarter of a million people obtained their water directly from ponds and rivers (Gorin 1968, p. 37). Not only was this a disgrace for the capital and administrative center of a country, but it would never do for what was intended to be one of the country's leading industrial areas. More water would have to be obtained elsewhere.

One of the first water supply projects authorized by the new Soviet state was the construction of a reservoir on the Istre River outside Moscow. Even this was not approved until the 1930s, but when it was completed, the flow of water to the Rublevsk station was increased from 9 to 14 cubic meters a second. As a result the daily water supply of the city almost doubled. The per capita availability of water also increased, but because of the rapid expansion of Moscow's population, per capita consumption increased about 50 percent.

The next stage of expansion was perhaps the most far-reaching. No matter how much they were dammed up,

the nearby streams simply did not possess the volume of
water necessary to supply Moscow. Consequently in 1933
work was begun on linking the Moscow River to the
Volga (Gorin 1968, p. 37). The completion of the Moscow–Volga Canal in 1937 coincided with the opening of
the Vostok (east) water treatment plant in the eastern
outskirts of Moscow (see Table 3.4). It took its water
from the Uchinsk Reservoir. This made possible almost
a doubling of the city's water supply, although again on
a per capita basis the increase was somewhat less. After
the war, work began on another water supply station, the
Sever, or northern, plant. Upon completion of the first
part of the Sever station in the early 1950s and the second
portion of the plant in 1957, there was again a substantial jump in the amount of water available. In 1955 it

Table 3.4 Water-Pumping Stations and Facilities in the Moscow Area

Station	Year Opened	1967 Capacity (1,000 cubic meters/day)
Rublevsk	1903	147
Cherepkovskaia Ochistnaia		53
Moscow-Volga Canal	1931–1937	
Vostok	1937	170
Sever	1948–1952 1957 finished 1971—third block	153
Zapad	1960–1966 1969	108 614

Source: D'iachkov 1967, p. 31; Chvanov 1971, p. 15; Gorin 1968, p. 36.

was decided to expand the capacity of all the plants by about 50 percent. To provide yet more water for the Rublevsk plant, another dam and reservoir were built at Marfin Brod. This made it possible to increase the flow from 14 to 22 cubic meters a second (Gorin 1968, p. 37).

Despite the increase in water supply, Moscow's population grew even faster. By 1960 the capacity of the existing stations was fully taxed. In 1964 it was reported that because the pumping stations were overloaded, only one-half of the city of Moscow had sufficient water pressure (Markizov 1964, p. 3). Therefore it was decided to build a fifth water supply unit in the western part of the city called Zapad (west). It was fully operational by 1966, and its water went to supply the new housing in the southwest region of the city. Until this time the west and southwest regions were continually short of water. By the mid 1960s new construction of housing in the southwest had diminished somewhat but had increased in the north. Not surprisingly, despite the existence of a city plan, by the late 1960s, the Sever water plant was overtaxed.

In the search for new water, new rivers must continually be tapped and new reservoirs must be built. The Ruz and Ozern rivers were diverted to the Moscow River in 1965–1966. The Ruz Reservoir is expected to hold 220 million cubic meters of water when full, and the Ozern Reservoir will hold 144 million cubic meters (Chvanov 1971, p. 15). These reservoirs are designed to feed into the Rublevsk and Zapad water stations. To provide additional water, it was decided to take water from the

Vazuza River (Chvanov 1971, p. 16; Gal'tsov 1964, p. 6). A large hydrotechnical system is now scheduled to open in 1976 (Chvanov 1971, p. 16; *Prav* 9/24/70, p. 2). The completion of the Vazuza system will provide Moscow with water from three sources, (1) the Volga via the Moscow Canal, (2) the Moscow River system, which collects water from the Istra, Mozhalsk, Ozuy, and Ruza rivers, and (3) the Vazuza (*Prav* 9/24/70, p. 2). The Vazuza system will require two dams. One dam, 883 meters long and 35 meters high, will be on the Vazuza River and will hold 539 million cubic meters of water. A second dam will also be erected on the Yauza River. It will be 1,073 meters long and 27 meters high and will hold 290 million cubic meters of water, of which less than half can be used (Chvanov 1971, p. 16).

  When it is completed, the Vazuza Reservoir will boost the capacity of Moscow's reservoirs to 2,000 million cubic meters (Chvanov 1971, p. 16). To do this, Moscow will have to reach out 180–200 kilometers from the city for water (D'iachkov 1967, p. 33). After that, the only other new source that can be tapped appears to be the Oka River, which may be needed by 1980 (*SN* 7/31/71, p. 207). Each time the water authorities must reach out farther and farther (Chvanov 1971, p. 16). But when a new area is tapped, the cost rises significantly and will continue to rise. During the Eighth Five-Year Plan of 1966–1970, Moscow spent 75.8 million rubles for capital investment on its water supply system (Chvanov 1971, p. 15). The Vazuza system alone will cost 128 million rubles. Bringing in the Volga River water costs 4.34

rubles per 1,000 cubic meters. This is 1.01 rubles per 1,000 cubic meters or 20 percent more than water that comes from the Moscow River (Markizov 1966, p. 12). The higher cost is due to the higher amortization rates necessitated by longer canals and the more expensive treatment that water from the Volga requires. The Volga Canal now supplies the city with two to three times more water than the entire flow of the Moscow River (*Sot In* 8/15/70, p. 2). The Russians are proud of this improvement and regard it as a technical accomplishment, even though the whole rerouting is expensive for them.

Another important source of water for Moscow as well as for other major cities such as Kiev, Tashkent, Baku, and Erevan is artesian wells. Factory managers find it particularly attractive to dig their own wells because the private cost of a cubic meter of artesian well water is about half the cost of tap water (Ianovskii 1964, p. 13).[2] Prior to 1959 there was virtually no restraint on such activity. Therefore many users, especially factories, availed themselves of this seemingly cheap supply. It was soon discovered, however, that the water table in the cities mentioned earlier had fallen about 40 to 60 meters and was continuing to drop 1.5 to 2 meters a year (Ianovskii 1964, p. 12; Grin and Koronkevich 1963, p. 120; *Kom Prav* 4/27/60, p. 2). A drop of similar magnitude was noted in several other cities including

[2] For the curious, water for both Moscow and Leningrad households and apartments costs four kopecks per cubic meter. The rate for industry is apparently twice that sum, although I was told the rate in Moscow was higher than it was in Leningrad. In both cities, the water administration makes about 60 percent profit.

Stravropol. In an effort to curb such activity, a special law "On the Strengthening of State Control over the Use of Underground Water and Measures to Protect Them" was passed on September 4, 1959 (Kolbasov 1965, p. 234). Except in emergencies no more underground water was to be withdrawn when other sources were available. Unfortunately, this law seems to have had little impact. In 1962 Moscow factories were still drawing about 260,000 cubic meters a day or about 20 percent of their total consumption from wells located within the city (Ianovskii 1964, p. 12). Because of the difference in the cost of well water and tap water, this was only to be expected. Undoubtedly the preference for well water will continue until something is done to embody the social value of artesian water in the cost the factory managers have to pay when they use this water. Not only has water withdrawal continued at about the old rate, but many factories discharge their industrial wastes, especially those that are particularly hard to handle, into underground wells (Bogever et al. 1966, pp. 27–28; *CDSP* 12/15/70, p. 17). This problem is serious in the Ostankino and Kazan regions of Moscow and the area near the Pavelets railroad station (Bogever et al. 1966, pp. 27–28; D'iachkov 1967, p. 34).

One way to alleviate the drop in the water table and the pollution of what remains is to stimulate the recycling of water. In the early 1960s only 12 percent of the water in Moscow was recirculated (Grin and Koronkevich 1963, p. 122). In the Ukraine, where water is in shorter supply, 18 percent of the water is reportedly recirculated;

but with the water shortages that have existed in both areas and the increasing cost of bringing in fresh water from ever greater distances, the relative proportion of recirculated water may soon begin to increase.

Despite the potential that should exist with a unified system of ownership and planning, not only is water that is difficult to treat not recirculated, but water that could easily be treated is reused only in limited amounts. In 1964 Moscow's electrical generating plants used 20,000 cubic meters of fresh water a day from the city water supply, but no effort was made to redirect this water for other industrial uses (Gal'tsov 1964, p. 8). In 1962 only sixty enterprises in Moscow plus some Teplovaia Elektrotsentral plants (TETs)—heat and electric power stations—bothered to recycle their water (Ianovskii 1964, p. 12). A. G. Ianovskii, a sanitary engineer, estimated that this number could be significantly increased so that 30 percent of all Moscow factories and almost all the steam and electrical generating plants could be converted. Similar potential exists in other cities including those in the water deficient Donbass region where industries have heretofore had a free hand to bloat themselves with potable water (Kolbasov 1965, pp. 53–55). According to Ianovskii, recirculating water would conserve about 70 percent of all the water used by industry in Moscow. Since about 48 percent of the water used by industry is needed only for cooling, recirculation should not prove too much of a problem. Such a change would significantly reduce the drain on the cities' freshwater supply and artesian wells, as Table 3.5 shows. On the basis of 1962 data, it is estimated that industry would take only 440,000 cubic meters of fresh water a day from the tap instead of 760,000 cubic meters. Since industry takes almost 40 percent of all the water consumed in Moscow, this step could measurably reduce the drain on the water system and the continual need to reach farther out for new and more expensive water supplies (Gal'tsov 1964, p. 7).

SEWAGE CONDITIONS Moscow's inland location also makes it difficult to find a place to dispose of water once it has been used. Obviously only the lower portion of the Moscow River can be utilized, since the upper drainage area must be carefully protected to ensure that it is safe to use as a source of drinking water. As a result of this, the western region of Moscow has been set aside as a sanitary water protection zone that extends 7,300 square kilometers (Drachev and Sinel'nikov 1968, p. 284). No industry can be located there. Instead it is an area of

Table 3.5 Sources of Water Used by Industry in 1962 and Potential with Greater Recycling (1,000 cubic meters per day)

	Present		Greater Recycling	
	Amount	Percentage of Total	Amount	Percentage of Total
City water supply	760	58	440	33
Artesian wells	260	20	115	9
Industrial water supply	120	9	430	33
Recycling	180	13	335	25

Source: Ianovskii 1964, pp. 12, 21.

farms, summer camps, resorts, and dachas for the government and professional elite. The sanitary protection zone is subdivided into three zones, with the strictest controls in force in the center. Regulations on the use of land are less rigid in the secondary and tertiary belts, but efforts are still made to prevent the discharge of sewage that might contaminate the water.

Conditions along the lower part of the Moscow River and toward the northeast, east, and southeast are not nearly as favorable. This is where all the sewage treatment plants are (see Map 3.1; Galanin 1967, p. 39). Because sewage discharge from Moscow will soon total 60 cubic meters a second, double the flow of the Moscow River, it appears that the downstream part of the stream could be more appropriately called the "Moscow Sewer" rather than the Moscow River (A. I. L'vovich 1963, p. 38). Moreover, until 1964 much of the sewage discharge was treated incompletely or not treated at all. There are some who worry that, even with biological treatment, there is not enough water available to dilute the sewage from the city (A. I. L'vovich 1963, p. 38; Fedenko 1966, p. 104).

Despite the inadequacy of the prerevolutionary sewage treatment facilities, the new communist government until the mid 1920s lacked the wherewithal to do much to improve the treatment of its sewage. In subsequent years when it had more resources, it apparently lacked much of a desire. Until the mid 1960s sewage treatment in Moscow, not to mention the rest of the country, was almost always subordinated to the task of increasing the

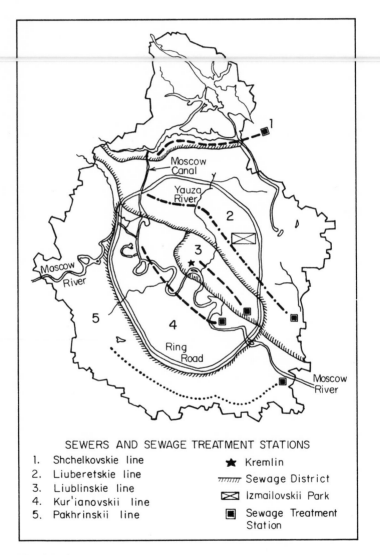

SEWERS AND SEWAGE TREATMENT STATIONS

1. Shchelkovskie line
2. Liuberetskie line
3. Liublinskie line
4. Kur'ianovskii line
5. Pakhrinskii line

★ Kremlin
🗥🗥 Sewage District
⊠ Izmailovskii Park
■ Sewage Treatment Station

Map 3.1 Moscow

freshwater supply. This can be seen in Table 3.3 by comparing Moscow's capacity to provide fresh water with its ability to treat its sewage. From the 1930s until 1965, the ratio of freshwater capacity to sewage treatment was either two or three to one.

The only sewage treatment facilities were at the Liuberetskie and Liublinskie sewage farms until 1929, when a third station with aeration facilities was opened at Kozhukhovskoi (Galanin 1967, p. 35). However, the rapid growth of the city and the failure to expand the city's sewer lines and sanitary treatment plants resulted in a serious water pollution crisis. A series of general plans were drawn up for the widening and reconstruction of the sewer system, but no effort was made to implement them until 1937. At that time, in addition to expanding the sewer networks, two aeration stations were built at Zakrestov and Filevskaia. The main work was the expansion of the Kozhukhovskoi station in late 1937 and the installation of aeration equipment at Liublinskie in 1938. All work was disrupted during World War II, but after the war new sewers were opened, and a new treatment plant was opened at Kur'ianovskii in 1950. The Kur'ianovskii plant was designed to treat the waste of the southwest region. Its operating capacity has been expanded greatly over the years, with major additions in the mid 1960s. By 1980 it will be capable of treating 3 million cubic meters per day of sewage. The Liuberetskie station will be similarly expanded to 2.5 million cubic meters per day and the Liublinskie station to 500 million cubic meters (Galanin 1967, p. 39). Plans have

also been drawn up for another sanitary plant at Pakhrinskii with a capacity for 2 million cubic meters which
will treat the wastes of the southern part of the city (see
Map 3.1).

Though there appears to have been a considerable
expansion of sewage treatment capacity in Moscow, there
is some question as to whether or not the sanitation
authorities have been following any long-range plan.
Sewage plants are often built and closed after as little as
a decade or so. In the list of Moscow's sewage plants that
have been built since the revolution, there are several
that are not expected to be in operation by 1980. In fact,
at the Sixth Scientific and Technical Conference on
Sewage Treatment in Moscow in 1968, a call was made
to close all but one of the smaller treatment plants now
in operation (*VST* 2/69, p. 41).

Russian cities have expanded so fast that housing construction has uncontrollably swept past what were once
isolated areas. In the Vyborg District of Leningrad, for
example, there is a sanitary treatment plant of the Riad
Grazdanskaia Prospect microregion. During a visit to
the plant in 1968, I found it was surrounded by the
50,000 people it was serving. The only water body
nearby was a little stream about ten feet across which
ultimately dragged its way into the Neva. Even though
the plant had been built as recently as 1957, the expectation was that it would be razed in 1973 and its operations transferred to a larger, more efficient plant that
was to built on the Gulf of Finland. In the meantime,
the population served by the Riad Grazdanskaia Pros-

pect plant is overtaxing its facilities. This shows a lack of foresight, disregard for capital expenditure, and a disinterest in the people who have moved into the area.

CURRENT EFFORTS TO IMPROVE MOSCOW'S SEWAGE TREATMENT With the consolidation and expansion of the main sewage treatment plants, Moscow officials hope by 1980, if not before, to terminate the dumping of untreated sewage into the Moscow River (*VST* 2/69, p. 41). Some authorities originally thought that this project would be completed by 1970, but there seems to be a difference of opinion about how successful the expansion of sewage treatment facilities has been. With the expansion of the Kur'ianovskii and Liuberetskie plants in 1963–1965, the amount of sewage dumped into the Moscow River was reduced from 540,000 cubic meters a day to 150,000 cubic meters (Markizov 1966, p. 4; Ianovskii 1964, p. 13; Ovsiannikov 1964, p. 17).

This expansion has done much to improve the quality of the Moscow, Likhororka, and Yauza rivers (Ianovskii 1964, p. 13). The manager of the Moscow-Oka River Basin Inspectorate for Water Utilization and Conservation, M. P. Yakolev, happily notes that fish have returned to the Moscow River near the Kremlin (*LG* 6/3/70, p. 11). Until 1965 the river at this point was practically devoid of oxygen. Other writers, however, seem much less sanguine. Expanded industrial production, especially by the chemical industry, has resulted in ever-growing industrial discharge, which seems to keep pace with or ahead of the completion of facilities for added industrial treatment. In 1969–1970, there were reports that 500,000 cubic meters of industrial sewage were still being dumped into the river and that many regions of the city still lacked sufficient sewer and water treatment (Chvanov 1971, pp. 15, 17). Even Yakolev acknowledges that conditions on most of the eastern tributaries of the Moscow River are still bad (*Trud* 11/12/66, p. 2).

With such intensive use of the water around Moscow, even good secondary treatment leaves the water in a sorry state. Despite the fact that about 6 million cubic meters of sediment and sludge are removed from the water in the process of treatment each year, the mineral content of the Moscow River continues to rise (*VST* 11/70, p. 37; A. I. L'vovich 1963, p. 38). During the summer of 1965, the river had a serious foam problem (Drachev and Sinel'nikov 1968, p. 286). Up to 0.1 percent of the water turned to foam as it flowed over the Sol-fino Dam in the Moscow River. This was due to the dumping of the detergents, nitrogen and phosphorous compounds, and oil products into the water.

Despite these unsatisfactory conditions, Moscow officials must be given credit for working vigorously to improve the quality of their fresh water and their sewage system. The city administration has allocated substantial sums of money for improvements. It is difficult to determine precisely how much has been spent. On the basis of scattered reports from 1959–1963, it is estimated that capital expenditures of 85 million rubles were made in supplying water and treating sewage (Markizov 1966, p. 4). In contrast, from 1966 to 1970, Moscow spent a total of about 190 million rubles.

As indicated in Table 3.6, there has been a relatively sharp increase in expenditures on sewage. For 1971–1975, expenditures of 400 million rubles are anticipated for both fresh water and sewage, while a longer-range projection for just sewage over the years 1968–1980 indicates that 430 million rubles will have to be spent (*VST* 2/69, p. 41; *LG* 6/3/70, p. 11).

It is possible that the growing cost of sewage treatment together with the growing cost of obtaining fresh water mentioned earlier will give fresh impetus to the policy of excluding new industry from the Moscow area (Markizov 1966, p. 12). In theory, no new industries are supposed to locate in Moscow, but, in fact, they do. Even though the full social costs involved in using water and air in highly congested areas are not reflected directly in the cost of production, the partial costs may reach a point where factory managers decide on their own to stay out of Moscow. After a time they are likely to conclude that the advantages of Moscow do not offset the disadvan-

Table 3.6 Capital Expenditures by Moscow on Water Supply and Sewage during the Eighth Five-Year Plan, 1966–1970, in millions of rubles.

Year	Total	Water Supply	Sewers
1966	33.2	14.6	18.6
1967	31.4	13.2	18.2
1968	32.4	14.4	18.0
1969	42.3	16.5	25.8
1970	51.0	17.1	33.9
Total	190.3	75.8	114.5

Source: Chvanov 1971, p. 15.

tages, which include among other drawbacks expensive air and water. As has happened in New York and is beginning to happen in Tokyo, Moscow's industry will be tempted to move to less congested areas. Ironically, it may take these market and operational forces to induce the implementation of long-range state plans.

Water and Sewage in Other Cities
The enormous effort being devoted to Moscow's water and sewage problems is far greater than anything else that is being done for other cities in the USSR. Moscow is the showpiece. Whereas other cities may be better located in terms of water supplies and sewage disposal, almost without exception such municipal needs have been attended to even later than was the case in Moscow. Even in Leningrad, the country's second-largest city, as recently as 1968 only 18 percent of the city's sewage was subjected to secondary or biological treatment, and only about 50 percent of it was provided with mechanical treatment and chlorination. Close to 50 percent of the water was to receive secondary treatment by 1971–1972, but it is uncertain whether these plans will be completed according to schedule (*SN* 2/1/69, p. 63). In the meantime, I was told in 1968 that Leningrad draws its water from the Neva at points upstream of its own sewage plants but downstream of other towns that also use the Neva for their sewer.

  For the country as a whole by January 1970, 1,736 out of more than 1,800 Soviet cities had at least some homes supplied centrally with water, while only 1,205 had sewers. By January 1971, 1,780 cities were expected to have centrally supplied water, and sewers were to be installed in a total of 1,250 cities (Ivanov 1970, p. 18). Two economists estimated that 88 per cent of all Soviet cities and 70 percent of the suburbs were equipped to provide running water, and 60 percent of the cities and 25 percent of the suburbs had sewers (Ivanchenko 1969, p. 233). These facilities were able to provide 50 million cubic meters a day, of which about 40 to 50 percent was subjected to treatment (Ivanov 1970, p. 18; Ivanchenko 1969, p. 239). Some republics fall considerably below the average. For example, in Georgia 56 percent of its sewage is untreated, and in Estonia 80 percent of it is untreated (*Sov Ros* 1/29/71, p. 1; Kolbasov 1965, p. 215).

Major cities and regions in the USSR find themselves with seriously inadequate water supplies and sewage treatment. The Donets and the Donbass regions are woefully short of water, and 536 million rubles will have to be spent to bring in water from other areas (Shabalin 1970, p. 12; Kaposhin 1971, p. 3). Moldavia and Belorussia are also water deficient areas. Minsk, the capital of the Belorussian Republic does not have adequate water supplies, nor does it have an operative sewage treatment plant (*Prav* 9/4/66, p. 2; *Sov Bel* 5/25/69, p. 2). Neither do Brest, Pinsk, and the other major cities in the republic. Conditions are no better in Moldavia. There two-thirds of the population in the rural areas and suburbs depends entirely on wells for the procurement of their water and outhouses for the disposal of their wastes. Of the twenty major cities in Moldavia, only six, Kishnev, Bel'tsakh,

Bendera, Tiraspol, Dubossar, and Rybnits have sewer
systems and of these only Dubossar and Rybnits, also have
treatment equipment (*Sov Mol* 6/11/69, p. 2). Other
major Soviet cities without adequate drinking water
include Sukhumi, Gagra (the resort areas of Georgia),
and Tiumen, Sverdlosk, Orenberg, Vladimir, Vorenezh,
Kharkov, Odessa, and almost all of the large cities in
Central Asia (*EG* 1/67, No. 4, p. 37; *EG* 5/69, No. 21,
p. 17; *Prav* 6/28/69, p. 3; *Prav* 8/27/70, p. 6; *Prav Uk*
8/21/62, p. 3; *Kom Prav* 4/27/60, p. 2; Ivanchenko 1969,
p. 243). Although almost all of these cities are located
near some body of water, that in itself is no guarantee
that water supplies will be adequate. For example, the
river Uvod' was polluted so badly near the city of Ivan-
ovo that it was necessary to build a canal 100 kilometers
long to bring in water from the town of Piles on the
Volga, a river not otherwise noted for the quality of its
drinking water (A. I. L'vovich 1963, p. 26). Similarly
the water in the mountain city of Kzyl-kul has become
a source of infectious diseases throughout the whole
republic of Kirgizia. Similar incidents have been reported
elsewhere in Central Asia (*Sov Kir* 9/17/70, p. 4; *Med
Gaz* 2/15/66, p. 1).

The existence of a central water supply and sewage
system does not necessarily mean that every home and
apartment is connected to it. Far from it. Up-to-date
figures are hard to find, but according to the census
figures of 1960, 62 percent of city housing in the USSR
lacked running water and 65 percent of the homes lacked
sewers (Kriazhev 1966, p. 130). More recent figures do

not show much improvement. Writing in 1969 the same two economists, V. I. Popov and Ia. M. Samoilov, who estimated that 88 percent of all Soviet cities and 70 percent of all worker's suburbs were equipped to provide running water, went on to say that only about 50 percent of the homes in these regions were actually equipped to receive the water. What was described as a "not large" (nebol'shaia) portion of them had sewage outlets (Ivanchenko 1969, p. 233).

Comparable figures for the United States may not exist, but roughly 70 to 75 percent of the American population obtains its water from central sources (U.S. Bureau of the Census, *Stat. Ab.* 1970, pp. 170–171). Of those who do not receive their water from a central pumping system, the overwhelming majority have electrically operated pumps in their homes. In the Soviet Union, such a convenience is virtually unheard of. As for sewers, over 70 percent of our population are served by some kind of central system (Goldman 1972, p. 22). However, about 10 percent of that sewage is discharged directly into our water courses as raw sewage and another 20 percent of it is given only primary treatment.

Most new apartment construction in the USSR is provided with water and sanitation facilities, especially that classified as being in the urban housing fund of the social sector (this essentially refers to apartment buildings as opposed to individual houses). (Makeenko 1968, p. 24; *Turk Is* 2/26/70, p. 3; *VST* 11/67, p. 4.) If we use this definition, 73 percent of the residential units in the

RSFSR have running water and about 70 percent are connected to sewers. Yet even by these standards, only 2.8 percent of the housing units on the *sovkhozy* (state farms) and only 1.4 percent of the *kolkhozy* (collective farms) have running water.

A trip through the Soviet countryside would seem to support the conclusion that there is a serious inadequacy of centrally supplied water and sewage. During a drive in 1968 from Leningrad in the northeast to Odessa in the south, virtually every village we passed along the way had open wells. Walking down one of the more fashionable streets of Kharkov, the fifth largest city in the USSR, or most other large Soviet cities including parts of Moscow, one sees many of the area's residents with yokes over their shoulders and a bucket at either end. This makes one skeptical of the assertion that as much as 50 percent of the housing units in the urban area have running water (*Prav Uk* 8/21/62, p. 3).

Nationwide Efforts to Improve Water Quality
The Russians have suffered some spectacular incidents of water pollution and environmental disruption, some of which are described in later chapters. Perhaps the most noteworthy was the outbreak of fire on the Iset River in Sverdlosk in 1965 and a similar incident on the Volga in 1970 (*CDSP* 12/21/66, p. 15; *Sot In* 7/4/70, p. 2). Like the Cuyahoga River in Cleveland, Ohio, which caught fire in 1969 and the Houston Ship Canal

which is a constant fire hazard, both Russian rivers were ignited when too much oil was dumped into them.

It has become harder and harder to find an untainted river in the USSR. Fish kills have occurred all over the country but especially in the major rivers such as the Volga, Kama, Oka, Kuban, Ural, North Dvina, and Dnepr (*Prav* 12/17/65, p. 6; *VST* 4/70, p. 11; Ivanchenko 1969, p. 239). One writer in *Pravda*, sarcastically describing what happened in the Ural River and its tributaries, said, "Today there is no fish problem, it's all solved." The fish had been eliminated! (*Prav* 1/10/68, p. 3.) Conditions have deteriorated so badly that in the Ukraine there is fear that the Dnester River may no longer be fit for use as a source of drinking water for cities such as Odessa. Reportedly almost all the major rivers in the Ukraine have lost their self-cleansing ability, so that there is almost no river in the republic which has been preserved in its natural state (Lysenko 1969, p. 57; *Rab Gaz* 12/15/67, p. 4).

While industry is not solely responsible for the deterioration that has taken place, it bears a good deal of the blame. Economists have estimated that 60 to 75 percent of all the industrial sludge in the USSR is not treated at all (Khachaturov 1969, p. 67; Ivanchenko 1969, p. 240; *EG* 1/67, No. 4, p. 37). For that matter, of the total sewage (industrial and household) that is processed by the country's sewage plants, only 20 percent of it receives secondary treatment (Zhukov 1970, p. 10). Comparable figures for the United States indicate that slightly over

40 percent of our sewage receives secondary treatment (Goldman 1972, p. 23).

Not only are Soviet treatment standards low, but very little effort is made to recycle water. This is to be expected as long as water is free or undervalued. Some writers have complained that some factories use four to six times more water than they need (*Kom Tad* 3/17/68, p. 3). This is reflected in the water-use norms that prevail in the USSR. As opposed to the two cubic meters of water required to produce a ton of steel in the Ruhr Valley (perhaps the lowest in the world) and the 100 cubic meters that is typical in the United States the Russian norm is supposed to be 115 cubic meters. In fact many Russian mills consume anywhere from 140 to 250 cubic meters per ton, although some mills have lower rates (Kaliuzhnyi et al. 1968, p. 218; Loiter 1967, p. 79; Armand 1966, p. 61; *EG* 1/67, No. 4, p. 38). Another study shows that one ton of oil can be produced with as little as 0.4 cubic meters of water. Several Soviet refiners use as much as 24 cubic meters per ton, again an irrational extravagence in many instances (Grin and Koronkevich 1963, p. 122; *LG* 7/8/70, p. 10). And as elsewhere in the world, the Russians are excitedly switching to synthetics, which generally necessitates the consumption of large quantities of water. Thus, just as in the United States, the production of one ton of synthetic fabric in the Soviet Union requires as much as 200 to 800 times more water than the production of one ton of cotton grown on unirrigated fields (Trofiuk and Gerasimov 1965, p. 55; M. I. L'vovich 1963a, p. 531; Commoner et al. 1971,

p. 4). Similarly, one ton of aluminum requires six to ten times more water than steel (B. S. Abramov 1970, p. 26).

Gradually the Russians like others around the world are coming to recognize the need to conserve and recycle their water. Notwithstanding the lack of a meaningful charge for the use of water, some industrial recycling is already taking place. The Russians seem particularly proud of the experiment being conducted at the Cherepovets Steel Mill. Here they recycle the air as well as the water (*SN* 9/14/71, p. 280).

Soviet hydrologists are also becoming increasingly aware of the impact of agriculture on the water supply. Although our main focus thus far has been on municipal and industrial use of water, it happens that agriculture is the largest consumer of water in the USSR. The Soviet economist M. Loiter estimates that agriculture consumes about three times as much water as industry. Moreover, whereas industry returns about two-thirds of what it uses to the sewer or river, farmers who take water for irrigation are one-time users of about four-fifths of what they take since the water either evaporates or sinks into the ground. The loss figure is especially high when water is poorly used or irrigation facilities are inadequately controlled (Ivanchenko 1969, p. 239). Finally an increased emphasis on the winter sowing of wheat has resulted in increased absorption of water by the crop and a decreased release of water to the surface. In some places this has meant a reduction of as much as 10 percent of the availability of water at the surface (M. I. L'vovich 1963a, p. 274).

"As Long as the Spring Flows, No One Asks the Price of Water"

This old Russian folk saying aptly explains the Soviet economist's dilemma (*EG* 8/18/65, No. 33, p. 15). No one cared about conservation of water when it was plentiful. Now, however, water is in short supply and therefore it is valuable. One way to reduce water consumption would be to charge for its use. As we saw earlier, ideology has frustrated those who have suggested this solution. For some Soviet theoreticians, the very idea that water meters should be used to measure water intake is a rather brazen idea (Kolbasov 1965, pp. 116–117; *Sel Zh* 6/30/70, p. 2). Nonetheless, some economists have continued to recommend this measure. For example, there was considerable debate as to whether to sanction a meaningful charge for water prior to the adoption of the December 1970 law on water conservation (*Sot In* 10/11/70, p. 2). Ideology, however, was too much to overcome.

As if ideology were not enough of a barrier to the introduction of a charge for water use, opponents of the idea have also been able to go back into Soviet history and show that when it was tried, such a system did not work. It turns out in spite of the ideological reasons for not having a water fee, such a fee was used at least twice before in the years following the revolution. But while ideology did not then prevent such a fee from being introduced, it shackled its effectiveness to such an extent that the whole concept was discredited. The first use of such a charge occurred during the 1920s and 1930s, but instead of being a deterrent to the use of water it almost served as a stimulus ("Otsenka prirodnykh resursov" 1968, p. 77). First of all, the fee was intended to cover only the delivery of water, not its value. Second, in order to stimulate cotton growing, the charge for water used in the irrigation of cotton was one-half what it was elsewhere for other products. Third, water was supplied free of charge to poor farms. Furthermore, the funds collected from such fees were designated solely for use on the expansion of the irrigation system not for an improved system of conservation or for the protection of the environment.

A second attempt to levy a water fee was begun in 1949. Once again poor farms were exempted and once again the fee was only set high enough to defray delivery costs, nothing more ("Otsenka prirodnykh resursov" 1968, p. 78). Unlike the previous experiment, however, collected funds were consigned to the general state budget, not the irrigation authorities. This was an improvement. Nevertheless for ideological reasons the fees had to be kept low. As explained to me by the economist M. Loiter, only a symbolic charge could be made. Moreover, the charge had been introduced shortly after World War II when economic conditions were very bad, especially in agriculture. To many, any fee that was imposed on agricultural activity was viewed as another way of taxing the peasant. The fee was not enough to stimulate the conservation of water, but just enough was collected to upset the peasants. Consequently, when Khrushchev came to power and decided to improve

economic conditions for the peasants, he began to provide subsidies instead of taxes ("Otsenka prirodnykh
resursov" 1968, p. 78; Kuznetsov 1970, p. 179). As a
result, the water fee was abolished in 1956. Subsequent
proponents of the concept have not been able to explain
away these unsuccessful first efforts.

Despite these setbacks, Soviet economists, including
such figures as Khachaturov and Federenko, have continued to advocate that a water charge be levied on all
industry and agriculture and in those cities where no
water charge is presently collected or where the existing
rate is unrealistically low (Khachaturov 1969, p. 67;
Gerasimov 1971, p. 208). For that matter, the 1970
Water Law does make provisions for a water charge
under certain conditions (*Sot In* 12/12/70, p. 2). The
Councils of Ministers in the various republics have the
power to initiate charges for the special use of water.
For example, even before the 1970 Water Law, all firms
in the Donbass, where, as we have seen, water is in short
supply, have had to pay about 0.3 of a kopeck per cubic
meter of water used. Although this is more than most
water users outside the major cities of the USSR have to
pay, it is still below the direct cost which is expended for
water (*Prav* 3/17/71, p. 3; Loiter 1967, p. 85). According
to Loiter, the Donbass charge was instituted in the 1950s
when an enlarged system of centrally supplied water was
authorized to supplement the inadequate natural supply
(Loiter 1967, p. 79). A similar charge is made in Kazakhstan where the Irtysh River has been diverted for use in
irrigation. Even Loiter, a sophisticated Soviet specialist

on these questions, was not entirely enthusiastic about
such charges, especially when they were imposed on the
farmer. Although his writing shows he is concerned about
water from the conservationist's point of view, in our
discussion of December 1970, he seemed to be reluctant
to tax the farmer with any further burdens. His desire
to improve the well-being of the peasant seemed to be
greater than his desire to increase the rational use of
water.

Regardless of why they were introduced, the proposals
advanced by Soviet economists show considerable
imagination. Some, including Loiter, insist that only a
marginal principle will work. If this principle were
adopted, the price of water would be differentiated by
region and river basin according to the relative scarcity
of water and the extent to which the effluent was polluted
(Mazanova 1969, p. 17; Khachaturov 1969, p. 74;
Loiter 1967, p. 83; Sukhotin 1967, p. 92; *Trud* 11/12/66,
p. 21). Some have even proposed complicated formulas
which they claim will make it possible to determine
precise water costs (Kuznetsov 1970, p. 180; "Otsenka
prirodnykh resursov" 1964, p. 75; Romanenko 1967,
pp. 42–43). Moving in the other direction, one proposal
by G. Astrakhantsev would apparently require every
freshwater user to pay three times as much for fresh water
as for a comparable amount of used water (*VE* 1/69, p.
107). He claims that the use of such a system at the
Riazan oil refinery resulted in a drop of 4 percent in the
amount of fresh water used. If used everywhere, however,
this would probably result in excessive expenditure on

water in some areas and possibly inadequate treatment in others.

The most intriguing and complex proposal has come from V. Shkatov. As he sees it, "It seems advisable first of all to establish payments for discharging industrial sewage into water bodies and exhaust gases into the atmosphere; moreover, these payments should be based on polarity principles. Thus for discharging a cubic meter of maximally polluted industrial sewage, payment could be exacted from the incentive fund (or from total profit left at the disposal of the enterprise) in the amount of three kopecks; for the discharge of a cubic meter of partially purified industrial sewage, one kopeck; for the discharge of a cubic meter of completely purified industrial sewage (on the basis of money received from enterprises discharging unpurified sewage), *a two kopeck bonus should be given to the enterprise*. The question of payments for discharging harmful gases, polluted air, and smoke should be approached in about the same way." (Shkatov 1969, p. 87.) He goes on to urge immediate adoption of this plan even if the precise value of natural resources has not been determined. Based on the relatively successful experience of the water control authorities in the German Ruhr where industrial and municipal water users must pay appropriate amounts for both the water they pipe in and pipe out, such precise measures may indeed be unnecessary. Shkatov, however, would have to face another problem with his unique scheme which the Germans do not have. If a bonus is paid for returning clean water, some Soviet plant managers might decide to go into the "not creating pollution" business. They could simply threaten to pollute and then generously abandon their plans when offered an appropriate incentive. This could turn out to be comparable to those gentleman farmers in the United States who give up farming to receive government bonuses for refraining from doing what these "gentlemen" were not doing anyway. There is a blackmail element here that might be hard to eliminate. Shkatov's proposal, nevertheless, suggests that some economists are very much aware of the underlying need to conserve water and are experimenting with some intriguing proposals.

The Cost

As elsewhere around the world, the implementation of the various proposals to improve the quality of water in the USSR is expensive. It is as difficult to establish what the costs are in the Soviet Union as it is in the United States (Goldman 1972, p. 19). The first point to establish is that there are two types of costs: those that arise out of the damage caused by pollution and those that must be incurred to eliminate the water pollution that already exists.

As might be expected, figures specifying the magnitude of damage caused by water pollution in the USSR are sketchy. Shkatov writes that the domestic fish industry loses 1.7 million centners (about 200,000 tons) of fish each year from pollution. This he feels costs the USSR 350 million rubles (Shkatov 1969, p. 68). Another economist, Astrakhantsev, calculates that "hundreds of millions

of rubles" a year are also lost from the failure to reuse all
the raw materials that are discharged along with the
sewage (*VE* 1/69, p. 107). At a conference in Alma Ata,
it was estimated that industrial water pollution alone
costs the USSR 1 billion rubles a year (a nice round
number) (*Kaz Prav* 11/26/67, p. 2). Finally, the agricultural economist B. Bogdanov figures that the cost is
actually several times higher. According to his estimates
the cost is 6 billion rubles a year from water pollution
and 4.5 billion rubles a year from water erosion (*SN*
4/14/70, p. 20). Comparable figures for the United States
have not been prepared by any responsible group, partly
because the definition of loss from water pollution is an
arbitrary concept.

Similar ambiguities arise when discussing the cost of
eliminating water pollution in the USSR. These discussions are complicated by the fact that the Russians
include the costs of irrigation and drainage work in the
total which naturally makes it look larger than it is.
Loiter estimates that the Soviet Union spends 700 million
rubles each year on repairs and new capital additions for
the country's water works. This estimate includes
drainage, irrigation, and hydroelectric operations
(Loiter 1967, pp. 76–77). He reported that in 1967 the
stock of fixed capital used for water management purposes (dams as well as municipal facilities) was worth
46.9 billion rubles. By 1975, capital investment in the
nation's water engineering system is expected to reach
50.1 billion rubles ("Otsenka prirodnykh resursov" 1968,
p. 69). According to another estimate it will cost 10

billion rubles in 1966–1970, or double what was spent in 1961–1965 to build reservoirs, sewers, and water pipes. The work on the Irtysh-Karaganda canal alone will cost 250–300 million rubles (Ivanchenko 1969, p. 233). Given these figures, it is not hard to believe that it might cost 17–20 billion rubles just to improve the quality of the country's water supply and sewage processing (Mazanova 1969, p. 17; Zhukov 1970, p. 12). If anything, Soviet estimates may be a little low. American specialists calculate that it will cost $16 billion to provide sewers and secondary treatment for those American households that do not have them. Control of industrial and thermal pollution will probably cost as much as $50 billion more (Goldman 1972, pp. 24–26). Because the Soviet sewage and treatment system is considerably less-developed, it is hard to see how improvement in the USSR would cost much less.

Conclusion

If we consider the conditions at the time of the revolution, the Soviet government has done an impressive job in upgrading its water supply and sewage treatment facilities. The years following the revolution, however, have also seen the expansion and the fulfillment of the industrial revolution within the USSR. Consequently, the magnitude and complexity of industrial and household effluent has also increased in a massive way. Presumably the increased output which has been generated by this growth in industrial output and capacity should also have been large enough to provide resources for financing the facilities needed to treat the increased flow of effluent. That is what one should expect in a planned economy. Still, based on the experience of the USSR to date, the Soviets do not seem to have been any more successful than most of the noncommunist industrial societies in coping with their increased effluent.

4 Air Pollution

By its nature, air pollution has tended to be a rather diffuse matter. Water pollution is a more tangible concern. If a particular water source is polluted, it is usually noticed immediately and it is generally not too hard to track down the source. In contrast air pollution, especially in a large city, is often a function of weather conditions, and it is frequently difficult to locate its origin.

Although air incidents in London in 1952 and Donora, Pennsylvania, in 1948 precipitated large numbers of premature deaths and alerted the world to the hazards of air pollution, somehow we tend to be less concerned about air quality than we are about water quality. In fact, until the late 1960s the symbol of economic growth throughout most of the world was dirty smoke pouring out of factory chimneys. The Russians have been no different in this respect than anyone else. Even now, such symbols periodically appear in Soviet literature.

Moscow

As with water quality, the history of air quality is closely linked with developments that took place in Moscow. Prior to the end of World War II, no one in Moscow or practically anywhere else in the USSR was concerned about the quality of the air. An air-measuring station in Gorky Park which opened sometime before 1945 was the only such unit in all of Moscow. In 1947 another station was added (Sokolovskii et al. 1965, p. 7). In 1949 the Council of Ministers of the USSR issued a resolution on air pollution and created the Chief Sanitary Epidemiological Administration (Glavnoe Sanitarno-Epidemiologicheskoe Upravlenie). (*LG* 8/9/67, p. 10.) From 1954 on, public health directors from the Ministry of Public Health and the Sanitary Epidemiological Administration organized a series of such observation stations so that by May 1955, there were five city and twenty-one regional observation stations (Sokolovskii et al. 1965, p. 7). It took some time to spread to other cities, but by 1966–1967, there were observation stations in sixty major Soviet cities (Nuttonson 1970b, p. 71).

Despite these observation stations, there was no institutional apparatus for dealing with air pollution until 1949. This did not mean that Moscow had no problems before that time. By resourceful use of data recorded for other purposes, air pollution officials have been able to ascertain that Moscow's air quality began to deteriorate significantly with the advent of the Five-Year Plans in 1930. By going back to the records compiled since 1870, naturalists were able to compare the average growth of pine trees during the years before 1930 with the growth since that time (Sokolovskii et al. 1965, p. 74). Pine trees are particularly susceptible to sulfur oxides in the air, and so pine tree growth can serve as an index of air quality. Records of such trees in Izmailovskii Park in the eastern part of Moscow were grouped by age and their average growth over a ten-year period of time were charted. The following changes were noted as shown in Table 4.1.

The rates of growth of the trees from 1930 fell sharply in some cases to as little as 10 percent of their former growth (Nuttonson 1970a, p. 20). Frequently many trees downwind of the main industrial areas simply died from the fumes in the air. Although not as severe, the same tendency has been noted in other less intensely industrialized areas of the city (Sokolovskii et al. 1965, p. 77). Tree growth in the Izmailovskii area was particularly affected however because industry in Moscow tends to be most heavily concentrated in the northeast and southeast parts of the city and the prevailing wind comes from the west and southwest. Consequently, from an ecological point of view, poorer air quality and the location of the sewage treatment plants in this part of the city make the northeast, east, and southeast regions generally less desirable places to live. This helps explain why the general public prefers to live in the southwest regions and why government officials build their dachas in the west and southwest suburbs.

The air pollution in Moscow and other large cities in the USSR was described in an article published in 1964 by two specialists from the Forestry Laboratory of Gosplan. "The growth of industrial establishments and

Table 4.1 Average Growth of Pine Trees in Moscow's Izmailovskii Park (in meters)

Age of Tree (years)	Average Growth in Height by 10-year periods	
	1870–1929	1930–1959
112	3.8	0.4
85	4.6	0.6
79	4.6	1.3

Source: Sokolovskii et al. 1965, p. 74.

large enterprises of factories and mills, as well as the development in the last few decades of automotive transport in Moscow, as in many other large cities in the USSR, caused the discharge into the air of millions of tons of gas, smoke, dust, and other pollutants. Accumulated smoke forming a dense, dirty smog can often be seen over large industrial cities. Quite frequently this dense, dirty fog hangs over the city at a height of 500 meters, and the ultraviolet rays have difficulty in penetrating it even in the heat of summer. In most cases, these rays do not reach the earth's surface." (Nuttonson 1970a, p. 13–14.)

At times the quality of the air in parts of Moscow has deteriorated to the point of affecting human health. The authors noted that "According to the data of many years of records of the Moscow Sanitary and Epidemiological Station, the maximum concentration of sulfur dioxide in one of the easterly suburbs (Izmailovskii Park) occurs in November–February, and the lowest in May–July. . . . Thus during May and June neither sulfur dioxide nor nitrogen oxides were found, whereas in October and November they reached a high level at all points, and at some points *their concentrations were near lethal.*" (Emphasis added.)

Recognizing the potential hazards in such a system, Moscow city officials moved vigorously. In what was probably the most ambitious program for environmental quality yet taken in the USSR, a determined effort was made to reduce the emission of harmful gases and particulate matter into the Moscow air basin. A series of steps were taken that probably will serve as a pattern for other cities in the USSR.

First, a determined effort was made to install smoke control equipment on the outlets of major emitters. Unfortunately the quality of the devices often left much to be desired, but the overall effort was decidedly positive. According to official data of the Moscow Sanitary Epidemiological Station, the number of installations with air purification equipment rose from 462 in December 1954 to 1,956 in December 1963 (Sokolovskii et al. 1965, p. 15). In 1960 alone, 137 units were installed at 36 enterprises.

Second, a number of the dirtiest factories were closed down or moved outside the city. From 1955 to 1964 twenty-nine factories and seventy-three shops were removed from the city (Sokolovskii et al. 1965, p. 16). Among these were several asphalt plants and a unit of Karpov, the chemical pharmaceutical plant which had been emitting manganese (Sokolovskii et al. 1965, p. 19). These measures alone brought about a reduction in the dust fall in the vicinity of the factory areas by as much as five to six times over what it had been in 1950.

Third, many individual home and factory furnaces and boilers were closed down and replaced with utilities that generated centrally supplied steam and hot water. These central heat and electricity plants, Teplovaia Elektrotsentral' (TETs), make it possible to provide such services with more efficient combustion.

Fourth, a decision was also made to upgrade the quality of fuel burned within the city. Until the mid 1950s, government planners did all they could to encourage Moscow fuel users to burn coal from the Moscow coal basin, which generally is of low quality. In addition many householders used lignite and peat, which were usually cheaper and more readily available. Peat and lignite, however, have an ash content of up to 50 percent and a sulfur content of from 2.5 to 4 percent (Sokolovskii et al. 1965, p. 39). When it was decided to bring in anthracite coal from the Donets basin there was an immediate improvement in the ambient air since some brands of Donets coal have an ash content as low as 4.5 per cent and a sulfur content of 1 to 2.5 percent.

Even more important than the switch to anthracite was the switch to oil and gas. In what appears to have been a serious planning and policy error, Soviet planners purposely discouraged the use of oil and gas in favor of coal despite the fact that the United States and Western Europe began to make such a shift soon after World War II. Oil prices in the USSR were kept high, and physical planning targets were drawn up so that they diverted oil and gas supplies elsewhere, usually outside the country. This decision, as much as anything, contributed to the continual deterioration of Moscow's air. Somewhat belatedly, therefore, the decision was made to substitute oil for coal and natural gas for both coal and oil. This led to an immediate reduction in ash. The reduction in sulfur emission was not as great because about 70 percent of Soviet oil has a heavy sulfur content, in some instances more than Moscow coal (Sokolovskii et al. 1965, p. 39). Moreover I was told by two Soviet engineers that it will take at least until 1974 or 1975 before the Russians have a desulfurization plant built to clean their oil. Natural gas, however, is low in both ash and sulfur, so the decision was made to use only gas wherever possible. Thus the flow of gas was increased in 1956 and 1957 and additional pipe lines were built and extended to other cities in the years that followed (Sokolovskii et al. 1965, p. 48).

Finally, the atomic test ban treaty brought about a noticeable reduction in the level of radioactive fallout. Reportedly, at no time before the test ban treaty in mid 1963 did the level of radioactivity in Moscow exceed the maximum permissible norms (Zykova, et al. 1970, pp. 50–53). In any case in the following years, the level of radioactive aerosols over Moscow fell by more than 95 percent and the concentration of other radioactive elements fell almost as much.

The cumulative effect of these various measures was noted almost immediately. Taking 1950 as 100 percent, the general level of ash emission fell to 46 percent in 1954 and to 21 percent in 1960 and 1961 (Sokolovskii et al. 1965, p. 43). The level of sulfur oxides fell somewhat less sharply, but it fell, nonetheless. In 1954 the level of sulfur oxide emission was 82 percent of 1950 and by 1961 it was 51 percent. In industrial zones, where the average annual reading of sulfur oxides was 0.81 milligrams per cubic meter in 1956, it had fallen to 0.24 milligrams per cubic meter by 1962 (Sokolovskii et al. 1965, p. 48). In residential areas the comparable decline was from 0.66 milligrams per cubic meter to 0.18 milligrams per cubic meter. On a citywide basis, sulfur oxide readings fell from an average annual level of 0.74 milligrams per cubic meter in 1956 to 0.25 milligrams per cubic meter in 1962. For comparison purposes, Philadelphia in 1962 had average annual readings of sulfur dioxide that were 0.23 milligrams per cubic meter and cities such as Cincinnati had comparable readings of 0.09 milligrams per cubic meter (CEQ 1971, p. 216). Particulate concentrations in Moscow in the same period fell from 0.81 milligrams per cubic meter in 1956 to 0.36 milligrams per cubic meter in 1962 (Nuttonson 1970a, p. 14). This compares with a maximum daily reading (a figure that is always much higher) of 0.13 milligrams per cubic meter in New York City in 1970 (*Business Week* 5/8/71, p. 18).

The radical improvement in the Moscow air from 1950 to the early 1960s did not continue at the same pace in the years that followed. One reason for this is that with a continuing expansion of the country's industrial base, more and more industries sought to locate in or near Moscow, including some particularly offensive ones (*LG* 3/5/66, p. 2). Khrushchev's emphasis on developing a chemical industry accelerated these tendencies. This more than offset the improvement due to the closings of other factories noted earlier. Thus, there are still traces of such gases as silicon, fluorides, lead, zinc, copper, chrome, and strontium in Moscow's air (Sokolovskii et al. 1965, p. 86). Some specialists insist that more factories should be moved outside the city limits and greater vigilance used to prevent new factories and apartments from sneaking into the green belt that surrounds the city (Sokolovskii et al. 1965, pp. 69, 89, 91).

If Moscow some day completely converts to a base of natural gas and electric power, it would probably still have an air pollution problem because of the growing size of its automotive fleet. Though many Russians have complained bitterly over the years about how hard it was to obtain a private car, at least they derived some comfort from the knowledge that a small automotive population brought with it a smaller amount of air pollution. It was with mixed emotions, therefore, that the air control authorities began to note the growing expansion of automobile and truck production. Whereas only 363,000 vehicles including 65,000 automobiles were assembled in 1950, by 1970 output had almost tripled, reaching 916,000 units including 344,000 automobiles (Tsentral'noe statistcheskoe upravlenie 1968, p. 197; *Sot In* 2/4/71, p. 1). Along major thoroughfares in Moscow at certain times of the day, there are already enough cars and trucks in the street to cause traffic jams. There are over 600,000 cars in Moscow, double the number in 1960, and the number of cars and trucks is due to increase rapidly after the completion of the Fiat automobile plant, the Kama River truck plant, and the reconstruction of some of the older auto factories (Sokolovskii et al. 1965, p. 28; *NYT* 5/6/62, p. 27; *NYT* 8/18/70, p. 14; *NYT* 10/31/71, p. 8). Automobile production alone should reach 700,000 to 800,000 a year by the mid 1970s, double the number in 1970. While this is still far below the 8 to 9 million autos produced annually in the

United States, it has already affected the quality of air in the larger Soviet cities. There is a promise of much more to come.

The first signs of automobile pollution were discovered as early as 1956 in a study by Z. V. Vol'fson and A. S. Lykova, who found that the average concentration of carbon monoxide in several large Russian cities was as high as 17–18 milligrams per cubic meter (Sokolovskii et al. 1965, pp. 29, 33). Subsequent tests in Tallin, Estonia, produced readings as high as 38 milligrams per cubic meter (Vasil'eva et al. 1970, pp. 95–99). This compares with maximum readings of 37 milligrams per cubic meter recorded in New York City in 1970 and an average annual level of 7 milligrams per cubic meter for Chicago in 1968 (CEQ 1971, p. 216). Conditions in parts of Moscow were not as bad, but even so, levels of 4.3 to 12.9 milligrams per cubic meter have been noted outside the city's center. Downtown, conditions are naturally considerably more serious. For example, readings of 70 milligrams per cubic meter have been recorded in Mayakovsky Square off Gorky Street. With the construction of pedestrian underpasses during the 1960s, traffic on Gorky Street is now able to move more smoothly. As a result, in the late 1960s the maximum concentration of carbon monoxide in Mayakovsky Square fell to 15 milligrams per cubic meter and the average concentration fell from 21 milligrams per cubic meter to 8 milligrams per cubic meter (Sokolovskii et al. 1965, p. 33). If we assume that the maximum concentration should not exceed 6 milligrams per cubic meter and the average daily concentration should not be more than 1 milligram per cubic meter, this was cause for concern (Sokolovskii et al. 1965, p. 92).

Because of their relative scarcity, automobiles in the USSR should have a smaller negative impact on the environment than they have in the United States. For a variety of reasons, however, Soviet automotive vehicles spew pollution far out of proportion to their numbers. First of all, because automobiles in the USSR have been in such short supply, the life expectancy of a Soviet automotive vehicle is much longer than an American vehicle. When repair costs become too high in the United States, it is much cheaper to buy a new car or a better used one. By contrast, in the USSR vehicles must last longer because it is usually impossible to trade in a used for a new vehicle unless one is an important official or has high priority on the waiting list. But the older a car or truck, the more likely it is that it needs a tuneup, which means that combustion is inefficient. In a recent survey, Academician Boris Stechken found that 85 percent of the motor vehicles in Moscow emitted three to five times the permitted amount of exhaust (*NYT* 11/7/68, p. 38).

Secondly, the quality of Soviet gas is very poor. Because there is a significant difference in price, most drivers use low octane gas without lead. In fact one authority claims that Moscow is the only city in the world where ethyl gasoline is banned (*CDSP* 3/18/67, p. 28). That is a bit of an exaggeration since Soviet gasoline with octane ratings of seventy or higher does contain lead. However, even if most Soviet vehicles use low octane gasoline, in the Soviet context, this does not necessarily mean there will be clean air. In most other countries of the world, the removal of lead is compensated for by the addition of aromatics and olefins which often produce their own forms of air pollution. Fortunately, the Russians do not fall into this gas trap. For the most part, their low octane gas has few additives. It is sold as it is. This eliminates one problem, but it causes a series of others, since without lead or other refining substitutes, the result is uniformly poor combustion. Thus combustion of brand A-56, a low-grade Soviet gasoline without lead, produces double the emission of carbon monoxide that results from using the same amount of another Soviet brand, A-70, with lead (Sokolovskii et al. 1965, p. 29). The fact that the engines that burn this fuel are likely to need tuning does not help matters. Nor is air quality improved by the Soviet practice of overloading their vehicles. This also leads to engine strain and poor combustion.

On top of everything else, Soviet gasoline and diesel fuel are high in sulfur content. Whereas the sulfur content of U.S. gasoline seldom exceeds 0.05 percent, in the Soviet Union the authorized standard is 0.15 percent. In diesel fuels the sulfur content runs even higher, to as much as 1.2 percent (Campbell 1968, pp. 170–171). In 1970 efforts were reportedly being made to remove a large percentage of the sulfur, but as indicated it is unlikely that these efforts will result in sulfur-free gasoline before 1974 or 1975 (*EG* 9/70, No. 36, p. 10). For that matter, the Russians do not appear to have made much progress

in installing an automobile exhaust inhibitor. Some
experiments have been conducted and some prototype
engines and catalytic devices have been built, but so far
the Russians have had nowhere near the success that the
Americans have had, and that success leaves a lot to be
desired (Gussak, p. 82; *SN* 9/23/69, p. 13; *LG* 9/6/66, p.
1; *Prav* 2/12/69, p. 1; *CDSP* 9/10/69, p. 34; *CDSP* 6/2/70,
p. 21; *NYT* 2/26/70, p. 6).

There are claims that photochemical smog is still a
rarity in the USSR (*Sov Kir* 12/22/68, p. 4). But this is
more a case of wishful thinking than of fact. There are
already reports in the USSR of policemen who have
become ill from too much exposure to the carbon
monoxide of automobile exhausts (*Sov Kir* 12/22/68,
p. 4). In the absence of higher grade gasolines and better
exhaust control, the main remedy so far has been to
increase octane levels in gasoline and speed up traffic
(Kazantsev 1967, p. 55). Vast sums are being spent on
pedestrian subways. (In the USSR, for the most part,
it is the pedestrian, not the automobile, that goes underground.) As we just saw, with cars moving faster, air
pollution readings in the immediate vicinity of the underpasses do show a significant decline in concentrations of
carbon monoxide and hydrocarbons (Sokolovskii et al.
1965, p. 33). But as the stock of automobiles continues
to increase, the concentrations have also started to rise
again. Moscow as well as other Soviet cities are finding
that such measures provide only temporary relief.

To some extent no matter what controls Moscow
imposes, it will always have air pollution. Partly because

of the heat-island effect caused by a large city like Moscow, it is highly susceptible to air inversions, particularly during the cold weather period. From September to April air inversions occur from 82 to 98 percent of the days and nights. Conditions improve a bit in June through August when only 66 to 70 percent of the days and nights are affected (Nuttonson 1969a, p. 93). That is why it is so necessary for Moscow to be as emission free as possible. And while Moscow has done more than most other Soviet cities, there remain serious cases of industrial air pollution. For instance the Karpov chemical and pharmaceutical plant, the Hammer and Sickle plant, and the Kleimuk factory as well as several of the TETs plants have been charged with polluting their neighbors (Kazantsev 1967, p. 55). Even if some day a program could be designed that would ensure the installation of more effective emission controls, there would still be no assurance that the problem would be eliminated. The continuing increase in industrial activity in metropolitan cities such as Moscow often offsets whatever improvements there may have been on a plant by plant basis.

During a flight to Moscow in December 1970, a gray blanket of industrial smoke could be observed covering the city up to its borders, where the blanket was sharply cut away. A few days later, Joseph Livchak and Sergei Kop'ev, two Soviet power engineers complained to me that there were still many areas in the USSR, including some in Moscow, where the air pollution norms were exceeded by two or three times what they should be. For

the most part the residents of these areas were not aware
that they were exposing themselves to serious safety
hazards.

Other Cities

The situation in most of the other large Soviet cities is
nowhere near as good as it is in Moscow. If air pollution
officials have found it difficult to prevent air quality
indices in Moscow from deteriorating, they cannot be
expected to have done much better elsewhere, particularly since the priorities outside of Moscow are almost
always lower. Indicative of the low priority generally
devoted to air pollution is the finding that throughout
the whole country only 14 per cent of the factories
acknowledged to be a source of air pollution have been
fully equipped and only 26 percent partially equipped
with air treatment facilities (Shkatov 1969, p. 68).

By concentrating their industrial activity and electric
and steam generating plants in large units, the Russians
do make possible better combustion possibilities. Although this may improve conditions in the city as a whole
because of the greater concentration of emissions from
one source, it may make air quality very poor in the
immediate vicinity of the factory or TETs. This is true
especially if funds and equipment for smoke control are
not provided promptly and amply. Since treatment
equipment is in such short supply, the effects of pollution
may be intense, expecially with the Russian practice of
locating factories in the midst of residential areas. In
cities with large metallurgical or petrochemical complexes,

the living conditions can be quite unpleasant (*LG*
8/9/67, p. 10; *Prav* 2/12/69, p. 1).

As is true elsewhere around the world, unfavorable
topography can intensify even simple air pollution. Like
Los Angeles, Tbilisi, the capital of the Georgian Republic, because of the arrangement of its hills, has air
inversions and smog 50 percent of the year (*Zar Vos*
5/23/68, p. 4). Also, like the metallurgical and chemical
mills in Pittsburgh, Birmingham, and along the Ohio
Valley, Soviet metallurgical, steel, and chemical plants
tend to be located in valleys, where natural ventilation
is cut off. This is where the ore or water supplies are usually located. Thus heavy-industrial cities such as Cheliabinsk, Sverdlovsk, Alma-Ata, Nizhnii Tagil, Magnitogorsk, Kuznetsk, and Karaganda have particularly dirty
air (Nuttonson 1970a, p. 23; *EG* 1/68, No. 4, p. 40; *LG*
8/9/67, p. 10; *Kaz Prav* 12/16/69, p. 4). Many of these
cities and others not located in valleys are continually
wrapped in a haze of industrial pollutants (Shitunov
1969, p. 82; *LG* 8/9/67, p. 10). In addition to industrial
effluent, there is also the automobile, and in at least one
instance, even leaf burning to contend with (*Kaz Prav*
12/16/69, p. 4). Indicative of the conditions were the
black snowfalls in Kaliningrad (*LG* 6/3/70, p. 11).
Because of the discharge of dirty air from its chimneys,
there are 40 percent fewer clear daylight hours in
Leningrad than in the rustic town of Pavlosk, only thirty-
five kilometers (twenty miles) away (Petrianov 1969,
p. 74).

Most air pollution is an inescapable by-product of

industrialization. There are many instances, however, where a bad situation has been intensified by poor planning and plant location. Though this happens frequently in noncommunist societies, presumably it should be a rare occurrence in a planned society. Nevertheless, industry in Odessa, for example, has been allowed to locate in the northern part of the city, despite the fact that the predominant wind comes from the north in both summer and winter (Akimovich and Ramenskii 1966, pp. 48–50). These winds carry the industrial fumes, as well as the dust storms, from the interior to all the resorts situated on the south side of the city. In Novokuznetsk, despite many protests from the city residents, an agglomeration plant was built after World War II in the single section of town with smoke-free air (*CDSP* 12/21/66, pp. 11–15). Subsequently, other metallurgical plants were added so that the town became completely surrounded with heavy smoke. In Gubakha, the "temporary" interweaving between residential and industrial sections became so permanent and unbearable that large residential sections of the town had to be relocated to another site at a cost of 20 million rubles (*CDSP* 12/21/66, p. 15). We saw in Chapter 2 how a similar planning error resulted in the construction and continual expansion of the Shchekino Chemical Combine upwind of Yasnaya Polyana, Tolstoy's former country home. Again the consequences could easily have been avoided with a little foresight.

On occasion, the nature of Soviet planning and the process of institutional organization and reorganization may actually frustrate efforts to ameliorate conditions. For example, in Cheliabinsk, despite initial attempts to locate industry outside of living areas, contractors were allowed to build new apartments near factories in order to reduce commuting time (*EG* 1/68, No. 4, p. 40). Now there is not one residential area not affected by some industrial activity. Smoke control is made more difficult by the widespread availability of coal. Air control authorities have been arguing that electricity generated with gas is cheaper and cleaner than that generated with coal or oil. Still, the coal combine in the region has insisted that coal be used to ensure a ready and accessible market for its output. As it is, doctors now refer to Cheliabinsk as a "metallurgical factory." The air is saturated with carbon, sulfur, and nitric oxides. In addition there are a zinc processing factory and a lacquer plant which spew out sulfur dioxide and titanium oxide in the middle of the city. Prior to the administrative reorganization of industry in late 1965, the Cheliabinsk Sovnarkhoz, which controlled and administered all industries in the region, had a plan worked out to recycle some of the wastes. The factories involved agreed to divert the sulfur gases from the lacquer factory to the zinc smelter, which needs sulfuric acid. When the Cheliabinsk Sovnarkhoz was abolished, and local control and coordination was transferred back to the reconstituted industrial ministries in Moscow, there was no one in Cheliabinsk with the power to implement this arrangement. The factories now belonged to two different ministries, both of which were located thousands of miles

away. Their respective ministries were not interested
in the nonproductive matter of air control.

Damages

From what we know about the effects of air pollution in
other countries, it should come as no surprise that air
pollution also takes a heavy material and human toll in
the Soviet Union. In one instance reported by the Soviet
specialist I. Petrianov, the social costs of industrial air
pollution were transformed into private or direct industrial costs. As they found more and more impurities in
the air, several metallurgical plants were forced to build
air tunnels several kilometers long in order to obtain the
unsoiled air they needed as an ingredient in their own
production processes (Petrianov 1969, p. 74). For manufacturers of electronic components, the need is not so
much for clean air as a part of the production mix but
as part of the working environment so that the products
will not be contaminated or their calibration thrown off.
In some Soviet production lines for radio components,
dirty air is said to be responsible for rejection rates that
are as high as 96 to 98 percent (Petrianov 1969, p. 74).

The devastation that can come from plant emissions is
suggested by what happened to the forests 20 kilometers
(12 miles) downwind of the magnesite brick factory at
Satke in the Urals. It, like the forest downwind of the
copper smelting plant at Copper Hill, Tennessee, has
been stripped bare (Armand 1966, p. 150). Similarly the
Karabash and Kyshtym forest preserves, which make up
the Zolataya Gora, have been completely deforested

(Nuttonson 1970a, p. 52). Thirty to sixty years ago the trees in these forests were intentionally cut to provide fuel, and now smoke from nearby factories frustrates all attempts to revegetate the area. Even in Minsk one of the local TETs plants emitted so much sulfur anhydride that there was a massive destruction of the pine forest in the Chelysukintsev Park, 5 kilometers (3 miles) away (Nuttonson 1970a, p. 43).

As in other countries, improper weather conditions combined with air pollution have caused serious incidents. Though Soviet public health authorities have never acknowledged the occurrence of anything as serious as the incident at Donora, Pennsylvania, severe air inversions and the emission of poisonous substances have occurred on many occasions in the Soviet Union. For instance a lacquer factory near Kiev was allowed to emit pure white lead from its stacks for several years (*Rab Gaz* 6/27/69, p. 4). Readings taken near the Central Ore Concentration Mill in Karaganda indicated levels of carbon monoxide as high as 1.33 milligrams per cubic meter (five to twenty times the maximum norm) and of sulfur dioxide from 0.28 to as high as 1.33 milligrams per cubic meter (four times the authorized maximum). On some days the dust concentration was as high as 4.7 *grams* per cubic meter (Nuttonson 1970a, p. 68). In Krasnoural'sk readings of sulfur dioxide and fluorides which are eight to ten times higher than the permissible standards are regularly recorded in the vicinity of the city's copper smelters (Nuttonson 1970a, p. 73). Until the early 1960s, the Polevskoy cryolite plant was spewing

fluorine into Polevskoy. Even after a fluorine retaining
unit was installed, so that the discharge was reduced to
~~25 to 33 percent of its former levels,~~ fluorine emissions
in the city continued to exceed the permissible norms
(Nuttonson 1970a, p. 3). Pathological work among
workers in the asbestos factory and residents of the city
of Asbest in the Sverklosk Oblast indicated a very high
rate of respiratory ailments (Ivanovskaia 1970, pp. 51–
52; *CDSP* 3/23/71, p. 26). (The same is true about
asbestos workers all over the world.) Similarly residents
within 2 kilometers (1.5 miles) of an electric power station
were found to be especially susceptible to disease. Of the
484 school children examined, 10 percent fewer had
normal lungs than a control group living 12 kilometers
(about 7 miles) away (Kaliuzhnyi et al. 1968, pp. 46,
49). The same pattern was noted among children living
two kilometers (1.2 miles) from a petroleum combine
(Balandina 1969, pp. 103–109). In the absence of official
confirmation it cannot be proved, but there is reason to
believe that the effects of such emissions have also
affected the population at large. With such contami-
nants in the air, the air inversions and poor circulation
patterns which were particularly serious in Sverdlosk
and Magnitogorsk in January 1967 and in Krasnouralsk
in the summer of 1960 probably produced a spurt in the
mortality figures similar to those recorded in London and
Donora (Nuttonson 1970a, p. 73; 1970b, pp. 2, 3).

Another aspect of the health problem that the Russians
are not too explicit about is the impact of radioactivity.
In time this may become an urgent concern as the Rus-

sians seek to increase the percentage of electric power produced by atomic power plants and the number of ships propelled by atomic engines. According to one authority, by the year 2000, the USSR will produce 2.13 $\times 10^6$ megawatts of atomic power and will discharge thousands of tons of radioactive waste (Medunin 1969, p. 31). Within the same period, the USSR will have commissioned about one thousand different atomic ships, some of which will also have to dispose of their waste in Soviet waters.

Although the Soviet citizen is officially told there is nothing to worry about, occasionally there are indications of deep concern. On December 19, 1968, the labor newspaper *Trud* published the following letter from one of its readers. "I live in Moscow not far from the Institute of Atomic Energy. They say that here there is a high level of radioactivity. This may explain why I am frequently sick. The radiation may be weakening my organisms. Tell me, are we threatened by radiation?" (*Trud* 12/19/68, p. 4.) Without hesitation, the editors of *Trud* asserted that there was nothing to fear but rumor itself. While they acknowledged that people who worked and lived nearby were sometimes worried about the hazards of atomic power plants, they insisted there was no cause for concern. As proof, the newspaper cited the large number of workers of the Institute of Atomic Energy who also live in the immediate vicinity of the Institute. In addition water wastes from Soviet atomic power plants are carefully treated and tall stacks are used to disperse airborne aerosols. Moreover despite reassuring promises that Soviet standards ensure that there will never be an accident, there have been some mishaps including several serious accidents in 1946–1948 when, among other things, many of the workers in an atomic plant developed cataracts of the eye. Similarly about 1963, despite the fact that officially it was claimed that the level of radioactivity in Moscow did not exceed the maximum permissible norms, there were rumors that radioactivity was causing a health problem in the capital. Reportedly, because of lax procedures in handling radioactive materials, there was a serious outbreak of leukemia in various parts of Moscow. In some neighborhoods, the disease was said to have reached near-epidemic proportions.

More recent studies indicate that workers in atomic power stations have not been subjected to abnormal doses of radioactivity, but the fear remains (Kozlov et al. 1970, pp. 54–56). Two Moscow engineers from the Steam and Electric Utilities explained to me in December 1970 that there was a rule that atomic TETs or power and steam plants must be located 20 kilometers (12 miles) away from the city for safety's sake. This presumably protects the city residents. Even then, the man in the Soviet street is not always happy about the construction of such plants. Not surprisingly, however, so far there have been no protests, or demonstrations against the location of a particular plant anywhere in the USSR.

Solutions

The techniques being used to reduce air pollution outside of Moscow are almost identical to those inside the capital. The chief method is to substitute gas for coal (*LG* 8/9/67, p. 10). In addition regional TETs operations have been set up. This has made it possible to shut down 300 boilers and furnaces in Cheliabinsk and 400 in Alma Ata (*EG* 1/68, No. 4, p. 140; *Kaz Pra* 12/16/69, p. 4). Efforts are also being made to install air filter equipment. Another approach has been to recycle some of the wastes (*Sot In* 10/20/70, p. 2; *Sot In* 10/23/70, p. 2; *EG* 10/69, No. 41, p. 37; Nuttonson 1969b, pp. 2, 3). Some success in both directions has been reported, but there are as many complaints about such efforts as there are accounts of success (Petrianov 1969, p. 73; *SN* 4/14/70, p. 20). One persistent difficulty is that there are more requests for equipment than there are facilities to produce or techniques to install the equipment (*LG* 8/9/67, p. 10; *Rab Gaz* 6/27/69, p. 4). One source says that only one-third of the requests for equipment can be satisfied, and others suggest that simply producing the equipment is not the same as ensuring that it will eliminate the air pollution it was designed to curb. Engineers have sometimes complained that they lacked the technical ability and resources to make such complicated equipment.

Surprisingly, another obstacle to designing better equipment has been that Soviet planners until the early 1970s were not able to provide the production arrangements necessary to produce the needed installations. By contrast, in nonsocialist societies pollution control equipment has become a big business. Since investors expect that there are big profits to be made, numerous private corporations in the United States specializing in pollution control research have found it easy to attract funds and set up their own producing arrangements. In the USSR, however, there has been little interest evidenced by existing or new state enterprises. Certainly new products and production techniques evolve from and are introduced periodically by the Soviet production system. Yet so far no industry or ministry in the USSR has apparently been willing to take the lead in promoting the innovation and production of pollution control equipment.

It is not easy to account for this lack of interest and enthusiasm. The explanation may lie in the peculiar nature of the Soviet economic system. Innovation and the transformation of laboratory ideas into the mass production process has been a major shortcoming in the Soviet Union. Difficult as it is to introduce products within existing industries and ministries, it is infinitely more difficult to arrange for the inauguration of an entirely new industry such as pollution control.

The belated introduction and expansion of the chemical industry is a good example of the kind of resistance involved. State planners from Gosplan and factory managers from other industries opposed almost all attempts to develop a major chemical industry in the USSR. Either they were reluctant to see capital investment funds diverted from their existing work, or they feared that the kind of innovation involved might jeopardize the fulfillment of their traditional plan targets. Ultimately Khrushchev himself had to take a hand and decree a nationwide campaign of chemicalization. Only by means of such an extraordinary effort could a notch be made for it in the established pattern of industrial activity.

Although it is not as unique and cohesive an industry as the chemical industry, the pollution control industry is beset with some of the same difficulties. Should a new ministry for it be created, or should the innovation and production required be assigned to an existing industry and ministry? If so, where does it fit in? Air and water pollution control equipment does not lend itself easily to categorization. As a result, it has taken a long time to to assign to a particular ministry the production respon- sibility for such activity (*Sot In* 8/20/70, p. 3). This is one area where the planned communist economy lags behind the uncoordinated noncommunist society. For a time air pollution control equipment was assigned to the Ministry of Oil Refining and Petrochemical Industries, but this ministry turned out to have too limited a jurisdiction (*LG* 8/9/67, p. 10; Sushkov 1969, p. 6). Until recently there was no research institute for the preparation of specialists and technical designers for dealing with various types of of emissions (Petrianov 1969, p. 79; *Sot In* 1/23/70, p. 3; *Prav* 3/24/69, p. 3). Complaints about the lack of research institutes and production facilities were heard from almost all of the republics (*Kom* 12/1/68, p. 4; *Sov Lit* 1/23/70, p. 2; *Prav Uk* 2/5/69, p. 2; 9/20/69, p. 1). The situation has been no better in water control (*VST* 2/71, p. 2; *Kom Prav* 7/24/70, p. 2). As a result, most of the installations that are ordered for control of air and water pollution must be custom made and custom installed. It seems as if every manufacturer must continually start from scratch and reinvent the air and water filter (Petrianov 1969, p. 79; *Sov Ros* 8/21/70, p. 2). Under the circumstances, maintenance has proved to be a serious problem, even by Soviet standards (Osipova et al. 1969, p. 39; *Rab Gaz* 6/27/69, p. 4; *Sot In* 8/20/70, p. 3).

Where technical design has not been an obstacle, financial difficulties have (Armand 1966, p. 151). Often insufficient money is allocated. What is allocated is often diverted for purposes other than pollution control or simply not spent (*Prav* 6/21/65, p. 2; *Prav Uk* 9/10/67, p. 3). As a result, actual installations of pollution control equipment are far below the number of requests for help received from factories that pollute. These requests in turn are far below the number of complaints and demands for action issued by government authorities (Sushkov 1969, p. 7; Lysenko 1969, p. 60).

When it seems that no way can be found to eliminate or neutralize air emissions, the authorities have occasionally ordered a plant to relocate or close down. We saw several examples of this in Moscow. In Riga, a superphosphate factory was closed, as was a lead paint shop in Krasnogorsk (*LG* 8/9/67, p. 10; *Sov Lat* 3/22/68, p. 2). Going a step further some public health officials have struggled valiantly to establish green belts as they close down or relocate factories (*Med Gaz* 3/15/66, p. 1). But on at least one occasion one critic noted that although fewer people were affected, the relocation of an

asphalt plant from Irkutsk to Slelekhov merely transferred the nuisance from one region to another (*CDSP*
10/9/68, p. 25). Similarly, rather than spend the money
needed to buy pollution control equipment, other
factories have closed down their inner city plants and
relocated directly in the green belt or sanitation zone.
The level of emissions continued as high as ever, but by
violating the whole concept and purpose of the green
belt the factories managed to obtain a greater dispersion
of their wastes (*LG* 8/9/67, p. 10).

It should not be implied that it is a simple matter to
close down or relocate a factory. First the laws are not
uniform among the various republics. In Uzbekistan,
Georgia, Azerbaijan, Lithuania, Kirgizia, and Estonia,
deliberate pollution of the air is not a criminal offense
(*Iz* 12/29/67, p. 3). Moreover, sometimes officials who
should have power to close a factory discover they are
helpless (*Sot In* 8/20/70, p. 3; *Prav* 6/2/65, p. 2). Even
public complaints from workers in the factory itself may
be of no avail (*EG* 3/68, No. 9, p. 20).

Cost

Estimates of how much it will cost to eliminate air
pollution in the USSR are even harder to find than
estimates of how much it will cost to clean up the water.
When Russian authorities want estimates of how much
is lost because of air pollution or how much it will cost
to clean it up, the usual habit is to refer to comparable
estimates made for the United States (Petrianov 1969,
p. 75). Included among the few figures the Russians have

published of their own is one for a dust collector for an open hearth furnace in the Ukraine, which costs about one and a half million rubles (*Rab Gaz* 5/17/68, p. 3). For the Russian Republic as a whole, T. S. Sushkov, a Russian lawyer, has calculated that expenditures on air pollution control totaled 155.6 million rubles during the years 1959–1967 (Sushkov 1969, p. 6). Undoubtedly, this does not include the cost of converting to natural gas. But if correct, Sushkov's estimate is considerably less than 10 percent of what we have spent in the United States over the same eight-year period for air pollution control. It also reflects the fact that except for the widespread conversion to natural gas, most Soviet efforts with air pollution control until recently have tended to focus on work conditions inside an individual factory or on particularly serious smoke nuisances in specific neighborhoods. If we note that most American specialists complain that the United States spends nowhere enough, the one thing that seems clear about air pollution control in the USSR is that the Russians will have to spend considerably more before they will be able to do much about improving the quality of their air.

5 Abuses of Land
and Raw Materials

Introduction

Until recently discussions about pollution have focused
either on the purity of air or of water. With very few
exceptions, air and water pollution were treated as
separate and essentially unrelated issues. This was also
reflected in the administrative approach to the problem.
In the United States, for many years water pollution
was the responsibility of the Department of the Interior
while air pollution was assigned to the Department of
Health, Education and Welfare. Only in 1970 was the
Environmental Protection Agency created in order to
handle both air and water pollution, as well as other
forms of environmental disruption in a coordinated
fashion. As we have seen, a similar fractionalization of
responsibility exists in the Soviet Union. As of this
writing, the Russians have not moved to coordinate
control with an equivalent to the Environmental
Protection Agency.

  Regardless of what the administrative response may be,
the fact remains that concern for the environment en-
compasses not only the interrelationship between air and
water pollution, but other uses and abuses of our natural
resources as well. How man treats the timber, land, and
raw materials at his disposal affects not only the quality
of the air and water but the whole ecological balance.
To ignore these broader aspects of the environment is
self-defeating in the long run. In addition to the Russians' handling of water and air pollution it is also

necessary to consider how they treat their other natural resources if we are to understand the overall complexity of environmental economics and conservation.

It is for such reasons that the concept "environmental disruption" seems a more appropriate description than "pollution." The introduction of any foreign element is a form of pollution. Whether or not this is something to worry about depends on how disruptive this foreign element may be. There is usually no harm, for example, in cutting down trees. After all, many trees are toppled by severe storms, but they are natural resources that potentially are renewable. The danger arises when bulldozers are brought in, the trees are cleared in a massive swath, and the earth is left bare. Then it may require all kinds of extra effort to stimulate new tree growth. The situation is compounded if no effort is made at replanting. Another example of the relativity of environmental disruption is the use of water that may be absolutely safe and tasty for human consumption but destructive when pumped into boilers that cannot tolerate certain minerals safe for humans. The term environmental disruption, therefore, implies that the dangers that arise from misusing the earth's resources are not limited to just air and water resources.

Remembering that environmental disruption exists in all countries and all societies, we shall, nevertheless, restrict ourselves to some of the more spectacular instances in the Soviet Union. Anecdotal as these cases may be, they illustrate some of the broad principles that guide the Soviets in the use of their land, timber, and other raw materials.

The Black Sea Shore

One of the most unusual examples of mismanaged land use in the USSR has occurred along the shores of the Black Sea. Because of its importance to the USSR, environmental disruption in this area is particularly unfortunate. Yet it is just because of its uniqueness that the area has been subjected to such intensive and not always well considered exploitation.

The Black Sea's significance is not hard to understand. For that matter any large water body in Russia's large land mass is automatically of special importance for recreational purposes, if nothing else. There is no other developed country in the world in which such a large percentage of its people live so far away from large water bodies. The residents of Leningrad are an exception, but, of course, it was for this very reason that the city of Leningrad was built.

Just by existing, therefore, the Black Sea would be important. Beyond that, however, the Black Sea is the warmest resort area in the USSR. It is not a Miami Beach and there are no coconut trees and alligators, but for the insular Russians who have traditionally found it difficult both politically and economically to vacation outside their borders, the Black Sea is regarded as the prime tourist region. There are more warm days there than at any other seaside resort in the USSR. Along its

shores are located such world-famous attractions as
Sochi, Yalta, Sukhumi, Odessa, and the newly built
resort strips of Mamaia in Rumania and the Golden
Sands near Varna in Bulgaria. Sochi was Stalin's cold
weather haven, and Khrushchev built a vacation home at
Cape Pitsunda, which is located about midway between
Sochi and Sukhumi. This magnificent peninsula is surrounded on three sides by mountains which are often
capped with snow. The mountains serve as a buffer from
some of the more rigorous storms of the continent. With
this protection, a magnificent strand of ancient and tall
pine trees has managed to thrive through the years and
provide the area with yet another enticement. Finally, its
temperate climate is congenial to the growing of citrus
products and early season fruits and vegetables.

Because it was so well endowed, Soviet planners were
encouraged to exploit the area's natural resources and
make it accessible to large numbers of Soviet vacationers
and government officials from the north. Not only was
one of the first jet airports of the country located at
Adler, midway between Sochi and Pitsunda, but a
massive construction program was inaugurated in order
to build hotel and resort accommodations along the
shores. [1]

Since much of the area on the east shore of the Black
Sea, as well as in the Crimea, consists of a narrow coastal

---

[1] The construction of Khrushchev's home at Cape Pitsunda (where he
played badminton on a Persian rug with Dean Rusk in 1963) explains
more than anything else why Adler had such a high priority as an airport
site. The Adler airport served jets in 1960 when republic capitals like
Erevan, Armenia, were still limited to propeller aircraft.

strip set off by a range of steep hills often extending to the shore itself, there is very little room on which to build and from which to draw materials. In some cases, hotels, roads, and a railroad bed have been built in spaces carved out of the hillside. To provide the concrete and other materials needed for construction, the contractors used the pebbles and sand located along the beach. Like the Riviera coastline, much of the Black Sea shore consists of small pebbles which would whet any cement maker's appetite. Because they were free for the taking and easily accessible and because obtaining other construction materials would necessitate the extra expense of transport over the mountains, local contractors used the beach materials.

While the geography was less restrictive in the west and northwest, contractors all along the Russian coast have mined the beach area since 1930. The resorts of Sochi, Tuapse, Gagra, Sukhumi, Poti, Batum, and Krasnodarsk were built largely out of beach materials (Zenkovich and Zhdanov 1960, p. 51). According to one estimate, 30 million cubic meters (40 million cubic yards) of sand and gravel were removed for construction from the Black Sea coast beaches in the postwar years (Zenkovich 1967, p. 27). In addition, the large seaports erected at Sochi and Yalta also were constructed with materials from the beach. Simultaneous with the construction of resorts, roads, railroads, seaports, and city buildings, the state expanded the electrical capacity of the region. This led to the construction of numerous dams along the mountain streams. Because of the proximity of the mountains to the coast, there was not much room for storage basins behind the dams. As a result, the dams were relatively small. Still they curbed the flow of silt and gravel from the mountains to the river mouths at the sea.

By the late 1930s, the combination of resort and urban construction along the coast and hydroelectric projects in the mountains began to alter the physical environment of the region. Until the mid 1930s, man's intrusion into the area, although dating back to pre-Grecian times, was limited, so that no serious disruption took place (Egorov 1962, p. 54). Gradually, thereafter, the beaches began to erode, so that by the 1960s, very few portions of the Sea of Azov and the Black Sea shore in the USSR had been left untouched.

In retrospect, it is easy to explain what happened. For centuries the pebbles of the beach had served as a buffer against the enormous power of the waves. At times the strength of the waves attained a force equivalent to 4 to 12 tons per square meter (Khodyrev 1964, p. 115; *Prav* 5/10/67, p. 2; *CDSP* 5/31/67, p. 29). Having been at Pitsunda during a winter storm, this author can verify that energy generated by the sea in this region can be violent. As the waves crash against the shore, they carry some of the shoreline with them (Khodyrev 1964, p. 115). As the contractors removed the pebbles they also removed a natural buffer essential for the prevention of beach erosion. Consequently, the sea rapidly ate away at the shoreline. When the ports at Sochi and Yalta were extended into the sea, the natural flow along the coastline was disrupted and the natural process of replenishing

the shore with pebbles and sand was disrupted (*Prav*
6/28/69, p. 3; Vendrov 1966, p. 16; Korzhenevskii et al.
1961, p. 60). At the same time, the newly built dams cut
off the supply of new pebbles from the mountain rivers
on their way to the sea. Previously these streams provided
about 90 percent of the sand and pebbles that ended up
on the beaches (*Prav* 6/28/69, p. 3).

The erosion that took place was often enormous. As of
1960, one specialist estimated that the beach area along
the Black Sea coast had shrunk on the average by 50
percent (Zenkovich and Zhdanov 1960, p. 51). The
shrinkage was due to both the contractors and nature.
What the contractors started, nature finished. Thus in
1914, the 100 kilometer (62 miles) spread of beach between Tuapse and Adler (including Sochi) occupied a
volume of 14 million cubic meters (18 million cubic
yards) of sand and pebbles. By 1956 this figure had been
reduced by one-half to 6.6 million cubic meters (Zenkovich and Zhdanov 1960, p. 53). Each year 320,000 cubic
meters (420,000 cubic yeards) of pebbles were removed
or swept away between these two cities and only 170,000
cubic meters (220,000 cubic yards) came in as replacement (Zenkovich and Zhdanov 1960, p. 53). Apparently
125,000 cubic meters of the disappearing 150,000 cubic
meters was removed by contractors for construction (*Iz*
3/8/67, p. 2). It is not surprising therefore that in the
intermediate stretch between Khosta and Adler what
used to be a 40-meter (44-yard) wide beach was completely eaten away by 1967. Near Adler, resort hotels,
port structures, hospitals and (of all things) the sani-

tarium of the Ministry of Defense collapsed as the shore-line gave way (*Prav* 5/10/67, p. 2; *CDSP* 5/31/67, p. 29). Particular fears that the mainline railroad, carrying 80 trains a day, would also be washed away have also been expressed. All of this was the inevitable result of man's thoughtlessness and nature's remorselessness.

Elsewhere along the eastern shore in places such as Krinshch at the mouth of the Pshad River, the beach which was 100 meters (109 yards) wide in 1950 had shrunk to 15–20 meters (16–22 yards) by 1960 (Egorov 1962, p. 55). At Pitsunda, after five modernistic fifteen-story hotels were built between the unique pine forest and the sea, an ill-considered effort was made to build a sea wall under the beach to protect the hotels. Along with everything else, this had the effect of unstabilizing the coast pattern. Soon 2.5 million cubic meters (3.3 million cubic yards) of beach material a year was being swept away (Zenkovich 1967, p. 47; *Iz* 3/8/67, p. 2). Professor V. Zenkovich, the head of the Beach Labora-tory of the Institute of Oceanography of the Academy of Sciences, predicted that this would eventually jeopardize the forest and the new hotels (*Iz* 3/8/67, p. 2). During a violent storm that swept the area in 1968, the sea swept past several of the hotel facilities and some of the build-ings began to sag. (The same thing almost happened again during my visit in December 1970.) Only by mobilizing all the trucks in the Autonomous Republic of Abkhazia in which Pitsunda is located and diverting them to the task of carrying in rocks and other solid fill were the hotels able to survive the inundation.

Similar happenings have occurred all along the coast. At Yalta, a beach that measured 28 meters (31 yards) in width in 1886 had shrunk to 15 meters (16 yards) by 1959 (Korzhenevskii et al. 1961, p. 60). Elsewhere in the Crimea, 8 to 16 meters of beach a year disappear (Panov and Mamykina 1961, p. 50). Because so much of the area along the sea is hilly, beach erosion often leads to landslides. In 1960, there were 300 landslides; in 1961, 342; and in 1968, 500. This can be hazardous not only for swimmers but for motorists. Four hundred meters of road between Yalta and Sudak just disappeared one day and another kilometer of highway in the area has also vanished. Even some streets of Yalta are threatened (*LG* 8/21/68, p. 11). The disappearance of the shoreline in some cases has also led to the contamination of freshwater lakes as salt water has swept in from the Black Sea (Dzens-Litovskii 1968, p. 64; *Iz* 5/15/70, p. 4).

The important port city of Odessa has also been plagued by the removal of beach materials (Zenkovich and Zhdanov 1960, p. 53). South of Odessa, contractors take away about 1 million cubic meters (1.5 million cubic yards) of beach material a year (*Iz* 3/8/67, p. 2). Much of the material has gone for dam construction. By 1963–1964 the continual removal of sand was blamed for some severe landslides along the coast (Vendrov 1966, p. 16). So much sand was removed or washed away from the estuary of the Dnester River that there was insufficient sand left to build a resort complex. At one time it was thought this project along the Dnester might compete with and complement the Golden Sands developments

280 kilometers (170 miles) further south in Rumania and Bulgaria. Apparently the Rumanians and Bulgars have taken more care of their prized beaches (*Iz* 3/8/67, p. 2).

These happenings did not pass unnoticed. A number of articles warning about the consequences appeared in various Soviet publications. Not only were scholars concerned, but the government itself devoted considerable attention to the crisis. Though a law banning the removal of pebbles from the beaches and feeder streams in Krasnodarsk Krae was passed in March 1962, the removal of the pebbles continued (Kazantsev 1967, p. 50). Consequently another, more inclusive law was passed in 1969 (*Prav* 2/26/69, p. 3). Little has happened to suggest that this second law was any more successful. On the contrary, the contractors seem just as active, if not more so, than before (*Prav* 6/28/69, p. 3; *Iz* 5/15/70, p. 4). Furthermore, a ruling that no new factories or buildings could be built within 3 kilometers (2 miles) of the coast has been ignored, apparently with impunity (*Prav* 2/28/69, p. 3).

Belatedly, large sums of money are being spent in an effort to restore a semblance of the natural balance to the area. From 1945 to 1960, the Ministry of Transportation spent 40 million rubles to strengthen the coast line, but to little avail. Some specialists have insisted that as much as three times that amount is needed (*Prav* 5/10/67, p. 2; *Iz* 3/8/67, p. 2). Gravel is being hauled in from inland mountains, giant cement slabs are being embedded in the sea coast, walls are being built, and man-sized cement blocks are being dumped along the

beach to replace with a buffer what has been washed
away (*Iz* 7/3/66, p. 5). Invariably the waves tear such
fortifications apart in six to eight years (Zenkovich and
Zhdanov 1960, p. 51). For some unexplained reason the
Russians until recently have not built many piers or
jetties perpendicular to the beach to deflect the tidal
force.

It is unlikely however that engineering measures alone
will suffice. As we have seen, a good part of the problem
stems from the fact that inadequate or no value is attached
to the gravel along the seashore. Since, in effect, it is
free, the contractors haul it away. As Professor Zenkovich
explained, "For the sake of temporary advantage, we
lose one of our most valuable natural resources." (*Iz*
3/8/67, p. 2.) Only when the gravel is priced high enough
so that the contractors conclude it will be cheaper for
them to go elsewhere will there be much chance of saving the beaches.

The United States also has to contend with such forms
of erosion. In New Jersey, 2,400 homes were swept into
the sea during the 1962 hurricane because they were
built too close to the sea. Fire Island off New York City
and the Gulf Coast off Louisiana are other locations
where excessive construction has generated beach erosion
which threatens to destroy whole islands. Still, the Black
Sea remains unique in that it is not only excessive construction which is causing the destruction but the actual
removal of the beach material by contractors who wish
to avail themselves of a "free" good. As explained in
Chapter 2, to some extent the absence of private property

rights may actually have hastened the deterioration of the Black Sea beaches. While the cost benefit calculation of a private landlord would probably not suffice to preserve a beach area from the lures of oil drillers, a private landlord might well consider his beach material to have a higher economic value (opportunity cost) when used as a resort than when converted into a rock and gravel quarry. The Russians cannot use such a first line of defense, and their Black Sea beaches have undoubtedly suffered because of it.

Kislovodsk

A different and perhaps more tragic example of the "free pebble effect" is occurring at Kislovodsk, high in the Caucasus. Here, too, failure to attach a proper (or any) value to a natural resource has brought about the degradation of a resort of national and even international fame. As along the Black Sea, the local contractors and planners saw what seemed to them a valuable resource virtually free for the taking. Regrettably they took it even though an opportunity cost calculation would have shown immediately that the resource involved had a higher social value as a mountain shield against the weather than as a quarry for limestone.

As before, some familiarity with the geography of Kislovodsk is needed in order to understand what is happening. Sheltered on three sides by mountains from the continental weather of the Russian land mass, Kislovodsk has long enjoyed a reputation as a warm weather oasis. Warm winds come wheeling in from the south

while the storms and clouds from the north have been
effectively blocked by the mountains. As one romantic
reporter put it, "It is as if the mountains were standing
guard to keep out the bad weather . . . which could only
squeak in by passing Kislovodsk's Thermopalae—a
narrow gorge between the two mountain ranges of
Dzhinal'sk and Borgustansk through which the Podkumok River flows." (*Iz* 7/3/66, p. 5.) Provided with
such protection, the meteorological records of Kislovodsk show that the city has had an average of 311 days
of sun a year while Piatagorsk, on the other side of the
mountains, has only 122 (*Iz* 7/3/66, p. 5). Reflecting the
abundance of fresh, clear, mountain atmosphere, one of
the attractions of the resort is called the Temple of the
Air (Khram Vozdukha).

Sometime after World War II, an unknown but enter-
prising bureaucrat from the railroad ministry strode into
this idyllic scene. His mission was to increase the volume
of railroad freight shipments in the area. He discovered
that the mountains and hills in the area were rich in
limestone. Without asking anyone, he arranged for the
construction of a lime kiln at the Podkumok railroad
station near the narrow gorge. "It was a small operation
and in the beginning nobody paid much attention to it.
When people finally did ask what was going on, it already
appeared to be too late to do anything about it. The
railroad and kiln operators met all arguments with, 'We
are a productive enterprise. Our product is sent all over.
We have an assignment and we are fulfilling our plan.'"
(*Iz* 7/3/66, p. 5.)

Sure enough, railroad shipments rose sharply as production increased. In the course of twenty years, seven additional kilns were put into operation. As explosion after explosion tore out the innards of the mountains for processing in the lime kilns, the residents of the area began to notice to their horror that what had been a narrow gorge was rapidly becoming a wide one. As along the Black Sea coast, the planners were so intent on extracting a raw material that they ignored the fact that they were also sacrificing a priceless heritage which "according to Leninist directive was transferred to the people as a whole—for their enjoyment and welfare." (*Iz* 7/3/66, p. 5.)

Now the miners are busily eating into the Borgustansk Mountains. Inevitably this has opened up Kislovodsk to the winter weather of the north. Combined with the dust from the lime kilns, the air of Kislovodsk is no longer unique. The dust level in the northern part of the resort now exceeds by one and a half times the dust norm allowed in a nonresort city. The critics see this as a tragic irony. On the one hand the state invests millions of rubles in new tourist facilities in Kislovodsk, while on the other hand the state destroys the very thing that makes the city so attractive. Moreover the destroyers are not only being paid a good salary for their vandalism but they are winning premiums for doing so in the name of "socialist competition."

Soviet conservationists find the barest satisfaction in blaming it all on "bureaucratic confusion." As one critic put it, "Logic is not obligatory in the functioning of a

bureaucracy." (*Iz* 7/3/66, p. 5.) There may also be some
compensation in finding that Marx was equally upset by
bureaucratic myopia. Despite such psychic compensation, however, the damage continues. The solution now
hangs on switching to another quarry so Podkumok can
be saved. Once an alternative supply of limestone is
obtained, then perhaps the Podkumok quarries will
cease taking the mountain apart and, in addition, bring
in other material to refill the original quarry. There are
also hopes that trees will be replanted and an effort made
to restore as much as possible of the former barrier. As
of 1966, however, after several years of anticipation,
production at Podkumok was increasing while that at the
substitute quarry was declining.

　Thoughtless exploitation of the area has affected more
than the air of the region. Kislovodsk, the Dzhinal'sk
Mountain Range, and the Narzan Valley are also famous
for Narzan, the Soviet Union's most popular mineral
water. The same area is also the source of Lake Tambukan mud which the Russians consider to have medicinal properties. Now, a combination of mining, timber
harvesting, and cattle grazing are threatening both the
mineral water and the mud. One mining crew was
delighted when they happened to tap an unexpected
gusher of mineral water in the Zheleznoi Mountains
(*Sot In* 8/16/70, p. 3). Their delight turned to dismay
when they realized that they were releasing all the pressure behind the Narzan mineral springs. Similarly the
removal of trees from nearby mountains and the grazing
of cattle on the hillsides threaten the premature drying

up of Lake Tambukan. It has been belatedly discovered that the mud of the lake originates on the slopes of the mountains. As it flows down, the mud and rainwater are normally saturated with mineral salts contained in the soil. The destruction of the forests has accelerated the flow of silt so that the mineral salts are now being washed out prematurely and possibly forever from the soil. Simultaneously the grazing cattle are packing down the soil in other areas and changing the composition so that the water passing over it is unable to absorb the proper salts (*Sot In* 2/28/70, p. 4; Chernyshev 1961, pp. 59–60).

Erosion

It has taken Soviet authorities a long time to appreciate the interrelationship between forest preservation and erosion. Although Soviet leaders, including Lenin and Stalin, have stressed the need for forest shelter belts, timber is usually regarded more as an obstacle to farming or a salable commodity than as an important link in the ecological chain. Like American pioneers, prerevolutionary Russian farmers as well as their kolkhoz and sovkhoz successors have wantonly removed all barriers in order to avail themselves of open fields. Equally destructive of the forests is the fact that timber has been a valuable earner of foreign exchange for more than a century. In 1969 for example, timber products constituted over 6 percent of all Soviet exports. If exports to hard-currency countries are taken separately, the figures are much higher. Thus timber constituted 20 percent of all exports to the United Kingdom and about one-third of

all Soviet exports to Japan. (Ministerstvo Vneshnei
Torgovli SSSR 1970, pp. 129, 255.)

Because of the demand for timber, it is only to be expected that Lake Tambukan is not the only water body
that has been adversely affected by a short-sighted
timber-cutting policy. As we shall see, the water in Lake
Baikal has been affected by the removal of trees from its
watershed as much as by the dumping of effluent from
the recently constructed paper and cellulose plants
(Chapter 6). Similarly the careless cutting of 25 percent
of the trees along the Black Sea in Abhazia has had a
deleterious effect on the water supply of the region (*Prav*
6/28/69, p. 3). What is bad for the water supply is almost
inevitably bad for the fish. The removal of trees leads to
a drop in the level of the area's ground water and an
increase in the flow of silt to the river. The increased silt
lowers the quality of the water and often smothers the
fish feeding areas (*Trud* 6/28/70, p. 2; *Sov Bel* 9/10/68,
p. 2; *Sot In* 8/1/70, p. 2). Then once the trees are cut,
they are generally tossed into the water so they will
float to the timber mills. Unfortunately much of the
timber rafted into the water in this way sinks to the
bottom where it absorbs oxygen and also adds to the silt
that is covering the feeding grounds (*Prav* 8/31/69, p. 2).

Timber cutting is hard to control. Traditionally the
peasants in Europe have raided private and state forests
to provide themselves with cheap fuel and construction
material. This as we saw was one of the things that upset
Lenin shortly after the revolution (Chapter 1). Despite
Lenin's fury, conditions have not improved much. Pro-

fessor Armand, the ecologist, complains that the areas within a radius of 30 to 40 kilometers (19–25 miles) of most large cities in the north and Siberia have been transformed into a wasteland as peasants and urban residents have chopped the trees down for heat and construction material. He expressed the hope that a law passed in August 1962 discouraging individual home construction and encouraging apartment buildings would reduce the destruction of the green belts around the cities (Armand 1966, p. 63). Presumably he must be worried once more because the draft Ninth Five-Year Plan of April 11, 1971, called for the resumption of support for private housing (*Prav* 4/11/71, p. 4).

It is not just the uncoordinated acts of individuals which are to blame. The pricing system in combination with the timber ministries bears a large part of the responsibility. Because of the absence of a good differential rent system, the land and timber located in populated areas are as cheap as land and timber in more remote areas. Under the circumstances, the timber authorities naturally select the more readily accessible timber, where labor and transportation costs are lower. As a result, 67 percent of all timber cut in the RSFSR comes from the European part of the republic, or the Urals, where now only 18 percent of the republic's forests are located. Siberia, which has 82 percent of the available forests is too remote, so only 31 percent of the timber is cut there (Sushkov 1969, p. 8). The situation is critical in the Ukraine where it is expected that by 1980 there will be no timber left to cut (*LG* 2/23/65, p. 2).

In addition to the massive timber cuts, the thoughtless ravaging of timber lands by the treads of Soviet tractors and bulldozers is a major cause of erosion (Armand 1966, p. 92; *Iz* 7/24/70, p. 3). Such practices have contributed to the landslides and beach erosion discussed earlier (Bulavin 1961, p. 59). Estimates are inaccurate, but some authorities speculate that as many as 50 million hectares of land have been affected by erosion (*LG* 2/25/70, p. 12; *SGRT* 3/62, No. 2, p. 18). Professor Armand calculates that from 1958 to 1964, 9 per cent of all the country's arable land has been eroded and 24 percent of the land actually under cultivation has suffered a similar fate (Armand 1966, pp. 24, 37). Others have estimated that erosion costs the Soviet Union 3.5 billion rubles a year (Eliseev and Kondratenko 1968, p. 114).

One of the more serious by-products of erosion has been the recurrence of dust storms in the USSR. As in the United States during the 1930s and again in 1971, this results from drought and the overcultivation of land in areas that have never been well endowed with water (*Kom Prav* 4/9/68, p. 4; *Kaz Prav* 3/23/71, p. 2). Some ecologists believe that overcultivation may sometimes actually help to bring on the drought and dust, which in turn reduce the crop potential of the area (Bryson 1971, p. 15). While dust storms predate the Soviet regime, such storms have increased in frequency since the collectivization of the late 1920s (Kravinko 1961, p. 67). This is probably due to both the increase in land put into cultivation and the Stalin-may-care attitude of the peasants who were never noted for their enthusiasm over collectivization. The Ukraine has been hit unusually hard. Dating from the collectivization drive of 1928–1930, it has had dust storms virtually every other year (Dolgullevich 1966, p. 34; *Priroda* 7/69, p. 124). Five oblasts in the Ukraine lost 1 million hectares of soil between 1950 and 1960. The city of Kherson has as many as ten dust storms a year and some cities in the Ukraine have reported as many as seventeen (Proshchenko 1965, p. 128). Khrushchev's decision to plow under the virgin lands unstabilized the soil cover and brought dust storms in the mid 1950s to Kazakhstan, which had been an area quiescent since the early 1930s when large-scale agriculture was discontinued (Zhukov 1970, pp. 33–34). When the storms throughout the southern Soviet Union increased in intensity in 1960, an ambitious plan was drawn up to reforest the region in order to stabilize the soil and temper the wind. Apparently neither this nor a subsequent plan announced in 1967 has been implemented with any success (Armand 1966, p. 105; *SN* 4/11/67, p. 24).

While other areas of Soviet land policy have not always had environmental consequences that were as disastrous as the virgin land program, economic development in the USSR generally has not been gentle with the land. Again this is due in large part to the underestimation of the value of the land caused by Soviet pricing policy. As explained by one Soviet critic, "Until now, land value has not been taken into account in determining the economic costs of construction alternatives." (*Prav* 3/10/70, p. 2; *CDSP* 4/7/70, p. 16.) Carried to an extreme, poor pricing policy facilitates enormous flooding of valuable agricultural land. As we shall see in Chapters 7 and 8, the failure to make allowance for the value land has as a producer of agricultural crops means that the dam acquisition and construction costs are seriously understated. Good land has been destroyed wastefully in other projects as well. There are increasing complaints, for example, about how too much agricultural land is being diverted to a host of nonagricultural uses such as warehouses or apartments. Such projects often do not yield as high a return as was the case when the land was used to grow crops. In an effort to prevent such irrational use of the land, one Soviet economist has suggested that prospective land users make what Western economists would call an opportunity cost calculation. If the land is presently being used as a meadow, then the yearly income that is currently being earned should be multiplied by 50; if it is being used as agricultural land, it should be multiplied by 100; and if it is being used as a vineyard, it should be multiplied by 200. If the land in its proposed new use is not worth as much as the figures obtained, then the land should continue in its original use ("Otsenka prirodnykh resursov" 1968, p. 52).

Strip mining is another illustration of what can happen when the pricing system does not reflect the full costs and benefits of a project. As is frequently the case, such distortions are not limited to the Soviet Union. Conservationists in the United States can find much that is familiar

to them in recent criticism about strip mining in the
USSR. Just as in Kentucky, many Russian naturalists
are now worried about the environmental consequences
of strip mining in the Kuzbass mining regions (*Trud*
6/28/70, p. 2). Initially the Russians were regarded as
backward for being so slow to engage in strip mining.
When in the mid 1960s they belatedly realized that strip
mining reduced direct mining costs, they introduced it
in a massive way. But as in so many other countries, the
Soviets have not yet included the social cost of strip
mining or the cost of reclamation in their cost benefit
calculations. Consequently the land has been ripped
apart without concern for the value of the land or the
environment or even for the agricultural contribution the
land could make if only it were restored. The latter at
least is a calculation made in many noncommunist
societies. When asked how much a hectare of Kuzbass
land cost, one engineer replied only partly in jest, "Two
pair of fashionable women's boots." (*Trud* 6/28/70, p. 2.)
Until recently the system has made no allowance for any
kind of official valuation, although some agricultural
specialists estimate that a hectare of land in the Kuzbass
might be worth anywhere from 20,000 rubles to 25,000
rubles when put to use in agriculture (*Trud* 6/28/70,
p. 2).

Solid Waste

After air and water pollution the next most serious concern of environmentalists is solid-waste disposal. But
because the Russians devote only minor attention to the

matter, solid-waste disposal will be treated here as part
of the broader problem of land and raw material utilization.

That the Russians do not concern themselves much with
solid-waste disposal is not surprising since they fail to
attach a proper value to land. Because land is treated as
a cheap or a free good, the Russians for the most part
use open dumps for the disposal of their solid waste.
Incinerators are rarely used. If it is decided to burn
before dumping, as is the case outside of some large
Soviet cities, the usual practice is to light a fire in the
open field. This leads to air pollution as well as the improper use of good land.

When it comes to recycling, however, the Russians are
more advanced than the United States in their treatment
of solid waste. Scrap metal, bottles, clothes, and paper
have a high value relative to labor in the USSR, so
there is a considerable economic gain to be made from
collecting such materials and turning them in for a fee.
Ironically, scrap materials in the USSR have a high
value, while Soviet land and raw materials within the
land have a low value. Because of depletion allowances,
just the opposite relationships prevail in the United
States. In our country, we worry more about the land
and its exploitable minerals; in the USSR they disregard
the land and worry more about collecting and recycling
what we would discard because the labor involved in
collecting and sorting is cheap. Consequently solid-waste
disposal is more of a headache here than there, where

less per capita is disposed and where almost any old field will serve as a dump.

Conclusion

While some uses and abuses of land resources in the Soviet Union are unique, the Russian experience is disturbingly familiar. Perhaps this is only to be expected as long as land values around the world do not incorporate social costs and benefits. To the extent that Soviet land values fail to reflect even private costs and benefits, however, the abuses of land in the Soviet Union will probably continue to be more extreme than they are in non-communist countries.

6 The Pollution
of Lake Baikal

The struggle of man against nature is a recurrent theme in much of the literary work of the past. As Ahab found in *Moby Dick*, more often than not man comes out second best. But that was in the era of the harpoon and the whaleboat. Today, with a vast new arsenal of technology at his disposal, man is much better armed. Seemingly the odds have been reversed in his favor. It is much easier now for man to impose his domination over nature. Unfortunately man's new triumphs generally turn out to be hollow victories.

A particularly depressing example of this modern-day struggle between environmental protectors and environmental disruptors is taking place on the shores of Lake Baikal in Siberia. The agony of Lake Baikal is symbolic of similar clashes all over the world. Yet the fate of Lake Baikal is more than a symbol. Baikal is a unique lake, and all mankind will suffer from its desecration. Nonetheless the struggle to preserve the purity of this wonder of the world runs smack against man's penchant to improve on nature. This is the conflict, and however honorable his intentions, man's efforts to "harness" Lake Baikal have been catastrophic. Perhaps the time has come to recognise that there are occasions when man must simply keep his hands off nature. These are the lessons of Lake Baikal. From both the dramatic and ecological point of view, man's assault of Lake Baikal has all the makings of a contemporary tragedy—man may indeed conquer nature, but in the process both nature *and* man lose.

The Lake

Because it has a sensual impact, Lake Baikal must be experienced to be appreciated. It is one of the few places in the world whose grandeur cannot be conveyed by words and pictures. Baikal is a lake of superlatives. Estimated to be 25–30 million years old, it is probably the oldest lake on earth (Lamakin 1965, p. 11). It is nourished by 336 rivers but drained by only one, the Angara, which about 2,500 miles later flows into the Arctic Sea (Rossolimo 1966, p. 61). Force-fed in this way, it would take 300 to 400 years to drain or replenish its water. This compares with 100 years to restock Lake Michigan (Galazii and Novoselov 1964, p. 3; Rossolimo 1966, p. 69). With a maximum depth of 5,346 feet, or just over a mile, it is the deepest lake in the world. [1] This exceeds the 4,700 feet of Lake Tanganyika, the second deepest lake on earth, and 1,932 feet of Crater Lake and 1,685 feet of Lake Tahoe, America's deepest lakes. Lake Baikal covers 12,200 square miles and contains 5,520 cubic miles of water, making it the largest body of fresh water on earth (*SL* 8/66, p. 6). This is about double the size of the next largest freshwater body, Lake Tanganyika, and more than twice as large as Lake Superior, the third largest lake. Lake Baikal contains about 2.5 percent of all the earth's fresh water, although some Russians claim that the figure is as high as 20 percent (*SL* 8/66, p. 6;

[1] At one time the lake was thought to be 1,940 and then 1,741 meters deep, but both of these estimates apparently were incorrect (Rossolimo 1966, pp. 32–33; *SGRT* 1960, No. 4, p. 84).

Sennikov 1968, pp. 16–18). The mineral content of the
water is 50 to 25 percent lower than that of most other
freshwater bodies (Sennikov 1968, p. 16). This unusual
purity is explained by the fact that most of the lake's
watershed is surfaced with rock, so that the water inflow
into the lake has little chemical or mineral contact
(Astrakhantsev and Pisarskii 1968, p. 73). As a result,
the water is highly transparent. Divers can see down to
a distance of almost 150 feet (Rossolimo 1966, p. 59).

The uniqueness of the lake is not limited to its water.
The climate near all water bodies is tempered because
of the temperature retention capability of water. With
so much water, the impact of Lake Baikal on the surrounding land mass is particularly noticeable. Since
large bodies of water are slow to change temperatures,
during the 30-mile drive from Irkutsk to the lake shore,
there is usually a temperature change of as much as 20
degrees Fahrenheit during the summer and winter. Even
though temperatures fall below freezing as early as
September or October, ice does not form on the lake until
late December or early January. It takes until May for the
ice to break up (Sennikov 1968, p. 18). The area is also
blessed with an unusually high percentage of cloudless
days. Even though the region is periodically subjected to
violent storms, Lake Baikal has an average of 2,583
sunny hours per year. This can be compared to a city
like Riga which has 1,839 or the onetime resort area of
Kislovodsk which has 2,007 hours (Sennikov 1968, p.
16). The quality of the air is also exceptional. As the

visitor approaches the lake shore, he risks suffocation as he tries to jam his lungs with the onrush of fresh, crisp air.

Reflecting the laboratorylike quality of the area, the flora and fauna of the region include species found nowhere else in the world. Over 1,200 living organisms have been catalogued in the area, of which 708 are peculiar to Lake Baikal (*Priroda* 11/65, p. 50). Included among these indigenous species are 30,000 nerpa, the world's only freshwater seals, and the golomyanka, a transparent fish whose young are born live, about 2,000 at a time.

So much about Lake Baikal is different from anything else man has encountered on earth that the lake is sometimes thought to have supernatural powers by those whose lives are touched by it. The area is venerated both by the local inhabitants and by Russian conservationists. But what is sacrosanct to the believer may be seductive to the nonbeliever. The lake is also seen as a voluptuous virgin whose resources are ripe for ravaging. All those fish, all those trees, all that pure water would whet the passion and set the juices running of any red-blooded promoter or industrialist. That the planner or manufacturer is a state employee in no way diminishes the hazard for the lake. Whether the enterprise be public or private, the richness and purity of the timber and water make it easy to expand production at lower explicit costs than would be the case in other areas of the country. It was only natural, therefore, that the area would sooner or later catch the fancy of officials in the timber, paper, and woodworking industries. If anything, the question is why did it take so long to decide on Lake Baikal as the ideal factory site?

Economic Development

Until the late 1950s, there had been no significant industrial development on the shores of the lake itself, which are mainly steep slopes unsuited for anything but small settlements. Several cities and factories, however, had sprung up on some of the lake's tributaries, especially the major one, the Selenga River, which supplies 51 percent of Lake Baikal's water. Over the years about 50 factories, including meat-packing plants and lumber mills, were established along its shores. Of these plants, only about 18 percent bothered to treat their waste before it was discharged raw into the Selenga (Astrakhantsev and Pisarskii 1968, pp. 102–103; *Prav* 12/9/68, p. 2). Most of these factories are located near Ulan Ude, the capital of the Buriat Republic. The city, which is about 75 miles from the lake's shore, also empties its sewage, untreated, into the Selenga.

Given the fragile balance of the region's ecology, it was only to be expected that the lake would show the effects of such encroachments by man. The catch of omul, the lake's prize fish, fell from 91,300 centners (about 10,000 tons) in 1945 to 40,197 centners in 1957, or by 55 percent (Buiantuev 1960, p. 22). To some extent it was due to the dumping of untreated sewage into the lake's tributaries, especially the Selenga which also happens to be the breeding area for 60 percent of the lake's omul (Buiantuev et al. 1962, p. 6).

The likelihood of much greater harm to the region increased, however, when in 1957 a fateful decision was made to bring industry directly to Lake Baikal (*Kom Prav* 5/11/66, p. 2). Plans were drawn up for the construction of a series of paper and pulp plants and lumber mills in the Lake Baikal basin. These plans were revealed by the economic planners in 1958, but not until July 1960 after work on the plants had begun was there any hint of what such industrial development might mean for Lake Baikal. Then in a small poorly circulated pamphlet which appeared in Ulan Ude, a local writer, B. R. Buiantuev, cautioned that plans for locating viscose cellulose plants and lumber mills on Lake Baikal would create complications for the lake and its timber cover (Buiantuev 1960, p. 38). How many people Buiantuev reached with his message is not known, but since his 48-page essay was published in remote Siberia in only 2,500 copies, it is unlikely that his cautions drew much attention. There were some other faint warnings emanating from Ulan Ude in 1961 in a book by two writers named O. Serova and S. Sarkisian, but again it is unlikely that many people outside this remote provincial city felt particularly concerned (Serova and Sarkisian 1961, p. 157). The public at large was unaware until almost 1962 of the plans for the pulp and cellulose mills or that they might be toxic for the lake. Then Gregory Galazii, the director of the Limnological Institute of the Academy of Sciences in Siberia, issued a public warning of what was happening to Lake Baikal in a letter to the editor of the mass circulation newspaper, *Komsomolskaia*

Pravda (*Kom Prav* 12/26/61, p. 4). He first described how
the plan was born. The Moscow office of the State Institute for the Design of Cellulose and Paper Plants (Gipp-
rosbum), under the stimulus of Gosplan (the state planning office) of the Russian Republic, suggested that all
those trees and that pure water be put to some use. Their
plans called for the construction of at least two plants,
one a cord cellulose factory on the lake shore at the river
Solzan, at what was soon to be called the Baikalsk plant,
and the other, a cellulose combine on the river Selenga,
the river's main tributary. After cataloguing the chemical
and biological changes such activities would inflict on
the lake, Galazii warned that the water waste from the
plants would not only destroy for eternity some of the
unique marine life of the lake, but it might adversely
affect the water supply of the city of Irkutsk. The waste
from the paper plants would flow out of the lake via the
Angara River which also served as the source of water
for Irkutsk.

  In what was to become a ritual plea for redemption in
the years to come, Galazii went on to suggest a number of
ways of averting the threat to the lake. First he urged
that the plants treat and recycle their water. If that
should prove to be too expensive, he suggested the alternative of building a 42-mile sewage bypass over the
mountains to the Irkut River (*LG* 1/29/66, p. 1). The
Irkut is the nearest suitable stream that does not flow
into Lake Baikal. Finally Galazii suggested that it was
not too late to abandon the whole project. He proposed
that the pulp and cellulose operation be transferred to

the city of Bratsk where the water quality was also good. Since a paper plant was already being constructed at Bratsk, all that had to be done was expand it. The water from the Angara at Bratsk was almost as good as when it left Lake Baikal, and there would be plenty of electric power available from the immense hydroelectric plant that would soon open nearby. Whatever happened, Galazii pleaded that no waste water from the cellulose plants be permitted to flow into Lake Baikal.

Except for another article a month later in the economic weekly, *Ekonomicheskaia Gazeta*, very little was heard about the impending danger for three more years until 1964 (*EG* 1/29/62, pp. 32–33). Then the debate exploded on the pages of another popular newspaper, *Literaturnaia Gazeta*, with a devastating attack by Oleg Volkov, who has since come to be known for his outspoken defense of Soviet nature. Other papers such as *Pravda, Izvestiia,* and *Komsomolskaia Pravda* soon joined in, but it has been *Literaturnaia Gazeta*, a literary newspaper, which in sporadic outbursts of frustration has continued to press the issue with the most persistence.

The Protest
Before proceeding to discuss the charges, responses, and results, it is important to say a word about the form such public protests must usually take in the USSR. Protests like those over Lake Baikal are, of necessity, protests by one government agency against the actions of another government organization. When a government news-paper decides to publish a letter to the editor or it

commissions a writer to publish such an attack, this usually indicates the existence of an interagency squabble. The public airing of such a fight almost always involves an embarrassing attack on someone or some organization. Actually such campaigns are quite common in the USSR and that is commendable. What remains unclear, however, is how the decision is made to launch such an attack. Many other instances of ecological destruction never find their way into the press. Moreover, there are no independent conservation groups like the Sierra Club or the League of Women Voters, which scrutinize the country like watchdogs looking for such abuses. Consequently when such debates emerge in the press, the implication is that the consequences are potentially far-reaching or that bureaucratic feuding has become particularly intense.

The role of the state in the exploitation of Lake Baikal points up an important lesson for those who argue that it is private enterprise that is at the root of the pollution problem in the United States. Certainly private business and private greed have done much to intensify the magnitude of environmental disruption in this country. However the elimination of private property is no guarantee that the environment will be any better off. After all, it was state officials who were directly responsible for the initial proposals to exploit Baikal's unmatched resources. From the state planner's point of view, there were easy profits and high-quality cellulose and paper to be made. As N. Chistiakov, the Vice President of the Ministry (then called State Committee) of Timber, Paper, and

Woodworking, and E. Kuznetsov, the Chief of the Cellulose, Paper, and Carton Administration of the Ministry, explained, "We are also for the preservation of the lake, but we are also opposed to underutilizing its huge wealth, its water, and timber." (*LG* 4/10/65, p. 2.) And as Chistiakov went on to ask me in December 1970, "Don't you believe in progress?" In other words, let's put the lake to work. But with a treasure like Lake Baikal, there remains the question of whether it is possible both to preserve the land and exploit its resources. If it is to be one or the other, it is not hard to predict on which side the officials in the Ministry of Timber, Paper, and Woodworking will be.

The republic and regional governors usually have much the same reaction. As we saw, under the system of five-year plans, with their specific production and profit goals, the most important measure of a successful administrator is whether or not he can show that production in his jurisdiction has increased. No quicker road to promotion exists. There is no place on their performance charts for grading a governor or mayor on his maintenance of the quality of the air or the water in his domain. All that counts is increased production and profits within their jurisdiction. Moreover the diversion of government funds in the Soviet Union to control environmental disruption is generally resisted by state and enterprise officials, since this usually means that there will be that much less money available in the region for expanding production. When asked why a new filter was not installed on a paper plant at Bratsk, one of the plant managers replied, "It's expensive. . . . The Ministry of Timber, Paper, and Woodworking is trying to invest as few funds as possible in the construction of paper and timber enterprises in order to make possible the attainment of good indices per ruble of capital investment. This index is being achieved by the refusal to build purification installations." (*CDSP* 3/24/65, p. 25.)

Both factory managers and political officials suffer if funds are diverted from *production* to *conservation*. In this respect, industry and political officials in the USSR have an identity of interests. Oddly enough, therefore, government officials in the USSR generally have a greater willingness to sacrifice their environment than government officials in a society with private enterprise where there is a degree of public accountability. There is virtually a political as well as an economic imperative to devour idle resources in the USSR.

While the emphasis on economic growth can also have disastrous consequences in a noncommunist economy, what is not appreciated here is that a politician in the United States is more likely to find himself in the middle as a mediator rather than on the side of industry. It is true that most American corporate officials would prefer to see less rather than more pollution control because such controls make production more costly and normally less profitable. Corporate officials have not been especially timid in making such sentiments known to government officials. But that is not all the American mayor or governor has to worry about. From the other direction they have to satisfy the Audubon Society and voter pressure groups such as the Council for a Livable World, Friends of the Earth, and local residents who suffer because of oil spills and polluted air and water. Of course, there are outspoken conservationists and suffering citizens in the USSR, and they are often supported by the Soviet press as we have seen, but for the most part they have neither the votes nor the clout, since what counts above all in the USSR is increased production. Moreover, the conflict in roles and interests seems to be unavoidable. The one major voluntary conservation group that is authorized in each republic is the Society for the Protection of Nature. The president of the branch in the Russian Republic, which has 19 million members, is Nikolai Ovsiannikov. He also happens to be the First Deputy Minister of the Ministry of Land Reclamation and Water Management of the RSFSR, the country's big dam builders and one of the major polluters in the country (*EG* 1/67, No. 4, p. 38). Understanding the reluctant role of Soviet officials in such circumstances helps explain why the attack on Lake Baikal has proceeded with such little effective opposition by political officials.

Indicative of the task facing those who now are trying to preserve Lake Baikal is the fact that a potentially effective law regulating the quality of Lake Baikal has actually been in existence since May 9, 1960 (Trofiuk and Gerasimov 1965, p. 58). Anticipating the difficulties the new pulp plant would create, the law—

1. Stipulated that the Selenga Cellulose Paper and the Baikalsk Cellulose Carton factories could not open until the factory officials could guarantee that the purification

systems were working and that all their effluent was
harmless. The plant could only open with the permission
of the State Sanitary Inspector (Buiantuev 1960,
p. 14).
2. Called for the enlargement of the Barguzin national
preserve around the shores of the lake (Buiantuev 1960,
p. 47; Trofiuk and Gerasimov 1965, p. 58).
3. Included a provision banning the continuous cutting
(or clearcutting) of timber when the slope of the land was
15 degrees or more and the selective cutting of timber
when the slopes were over 25 degrees. In addition there
was to be a program of forest restoration. The timber
bans were needed to prevent the erosion of soil covering
in the mountainous watershed of the lake.

All three provisions were ignored. The passage of a law
does not mean that it will be enforced, especially when
the interests of the governing officials do not coincide
with the intent of the law (*SL* 8/70, p. 54; *Kom Prav*
2/11/69, p. 1; *Iz* 2/8/69, p. 2; Goldman 1967, p. 160).

As it became clear that no one was paying any attention
to the law, the public outcry that followed led to the
formation of various investigative commissions. It is
hard to total up all the commissions that investigated the
issue at one time or another, but public reference has
been made to at least four major national groups that
opposed the opening of the paper plants and six smaller,
mainly local organizations that approved it.

Among those that disapproved of the plans for the
factories:
1. The Interdepartmental Commission of the East

Siberian Academy of Sciences, December 22, 1961,
February 24, 1965.

2. The Expert Commission of the Government Committee for the Coordination of Scientific Research,
February 23, 1962.

3. The Geographical Society of the USSR, March 1962
and March 1965.

4. The Academy of Sciences of the USSR, March 29,
1966.

Those whose decision is unknown:

1. The Expert Commission of Gosplan of the USSR,
March 15, 1966.

Those which approved of the plans at least at one time:

1. The Eastern Affiliate of the Siberian Academy of
Sciences, January 7, 1964.

2. The Irkutsk Oblast Sanitary Epidemiological Station,
December 17, 1963.

3. The Technical Council of the All Union Sewer Design
Bureau, March 17, 1964.

4. The Chief Administration for Fish Control of the
Ministry of Fisheries, April 14, 1964.

5. The Buriat Sanitary Inspector of the Administration
of Land Reclamation and Water Management of the
Council of Ministers of the Buriat Republic.

(*LG* 4/13/65, p. 2; *LG* 6/2/66, p. 2; *Kom Prav* 5/11/66,
p. 2; *Kom Prav* 6/9/66, p. 4.)

As has occasionally happened elsewhere around the
world, the testimony of at least one of the approving
experts was tainted when it was brought to light that he
had served the Baikalsk plant as a consultant for the

Ministry of Pulp, Paper, and Woodworking for a fee of 100,000 rubles a year for his laboratory (*Kom Prav* 8/11/70, p. 4).

Yet as the commissions labored, so did the construction workers. The commission reports fell off the presses like so many drops of effluent from the Baikal's paper plants —but unlike the effluent in Lake Baikal, the reports had no effect.

Despite all the laws and all the protests, it was readily apparent that no one was willing to act. It was a game of bureaucratic bluff; everyone agreed that the lake was being threatened, but all pleaded helplessness. For example, when the fish catch diminished, the fishing authorities complained that they had no authority to prohibit the discharge of sewage into the water. Only the Sanitary Inspector could do that, they said, and, for whatever reason, the Sanitary Inspector did not do it (*Prav* 2/28/65, p. 4). As for the waste from the paper plants, authority over such discharge was vested in the Ministry of Timber, Paper, and Woodworking and not in the Ministry of Land Reclamation and Water Management (*LG* 10/11/67, p. 12). That was like putting the Atomic Energy Commission in charge of regulating the radioactive discharge from atomic energy plants. One group of critics predicted that the Ministry of Timber, Paper, and Woodworking would end up transferring the responsibility for the discharge of waste into the lake to the paper enterprise itself (*Kom Prav* 5/11/66, p. 2). Such confusion and buck passing made it all the easier for the pulp and paper plant officials to churn cheerfully and reassuringly ahead. All that suffered was Lake Baikal.

Implicit in all this is the suggestion that the planners and officials in the Ministry of Timber, Paper, and Woodworking were something less than vigilant in their use of Lake Baikal. Certainly they uttered the appropriate expressions of concern, but there is little evidence that they exercised due diligence in their treatment of the lake.

Mistakes

A recurrent criticism by the conservationists was that the plants at Baikalsk and Selenga could have perfectly well been built elsewhere. Initially the Ministry of Timber, Paper, and Woodworking had insisted that the two plants had to "use Baikal's exceptional ultrapure water because it was impossible to make super, super cellulose cord" so essential for aircraft tires without it (*Kom Prav* 5/11/66, p. 2; *LG* 2/10/66, p. 1). As it turned out, these plants were built under false pretenses. Challenging the whole rationale for building the Baikalsk plant, the editors of *Komsomolskaia Pravda* noted that after the plant was well under construction, the Ministry decided to change the product mix. Somewhere along the way, the dire need for "super, super cellulose" was downgraded, and the Ministry decided to switch part of the plant's capacity to the production of paper, a product that does not require such high purity water. "In this way the original insistence on the need to locate the plant directly on Lake Baikal underwent a qualitative metamorphosis," the editors pointed out. "And what remains?" they asked.

"A nearly completed factory that with a little scientific effort could have been located in another region of the country and thus could have spared Lake Baikal." (*Kom Prav* 5/11/66, p. 2.) The production plan for the Selenga plant underwent a similar change. Once construction had started, the Ministry decided the plant should also produce heavy cartons which in no way required anything as pure as the water from Lake Baikal (*LG* 2/10/66, p. 1; *LG* 4/15/65, p. 2; *Prav* 2/28/65, p. 4; Trofiuk and Gerasimov 1965, p. 53). At the least, said the critics, the two plants should have been combined into one that could concentrate on "super, super cellulose" production. The elimination of the other plant would have made it a little easier on the lake.

As if all this juggling were not enough, the critics of the Ministry of Timber, Paper, and Woodworking next charged that it was foolish to seek self-sufficiency in the production of "super, super cellulose cord" in the first place. They assert that "super, super cellulose" has been superceded by nylon cord, which is a superior material. As proof they point out that factories producing "super, super cellulose" in Canada have closed down for lack of a market (*LG* 4/10/65, p. 2; *LG* 4/13/65, p. 2). Another obvious question is how is it possible to produce "super, super cord" in countries such as the United States when American factories do not have water comparable to that in Lake Baikal? Clearly less pure water will do. Such arguments have apparently made little headway: In fact as recently as August 23, 1970, the *New York Times* carried an article in which Russian officials indirectly blamed the United States for the pollution of Lake Baikal. The Russians were quoted as saying that they had to build their cellulose plants on Lake Baikal because the United States is the only other country in the world that makes such strong cellulose, and it refuses to sell it to the USSR. To the conservationist, such statements muddied the waters even more.

The Ministry authorities also adopted a self-serving stance when they tried to demonstrate the economic feasibility of their plans. Less polite critics insist that the Ministry purposely understated construction costs by as much as 22 million rubles, or by one-third of the total cost, in order to win initial approval for the plant (*LG* 4/15/65, p. 2). Future production costs were similarly minimized. For example, the planners failed to ascertain whether or not the supply of timber would be adequate for more than 25 years. Also overlooked was the fact that Baikalsk was located 60 miles from its timber supply base. Erratic deliveries of timber supplies have already forced the closing of the plant's production lines several times (*LG* 10/11/67, p. 12). Moreover the cellulose produced at Baikalsk required further processing at another plant located thousands of miles away in the Urals, where the cellulose plant could have been located in the first place (*LG* 2/6/65, p. 1; Buiantuev 1960, pp. 38–40).

Added to the increased cost was the somewhat belated recognition that the entire Baikal region is in a very active seismic zone (*LG* 4/15/65, p. 2; *LG* 2/10/66, p. 1). The site chosen for the Baikalsk plant turns out to be directly on the fault of one zone (*Kom Prav* 6/9/66, p. 4). The

Selenga plant has a somewhat better location. It is 16 yards away from the epicenter of a quake recorded a few years ago (*LG* 2/10/66, p. 1). Some reports warn that both plants could disappear into the lake. Major earthquakes occurred in 1862, 1950, 1957, and 1959 (Buiantuev et al. 1962, p. 9). Smaller shocks occurred in 1960 and 1961. During the quake of 1862, eighty square miles (about the size of Boston) of dry land suddenly collapsed into the lake (Galazii and Novoselov 1964, p. 4). Apparently in their early planning, the Ministry of Timber, Paper, and Woodworking completely ignored the likelihood of future earthquakes in the area and what the destruction of the factories in such a quake would do to the lake's water (*Kom Prav* 6/9/66, p. 4).

Throughout the entire debate, officials from the Ministry of Timber, Paper, and Woodworking insisted that whatever the plants would be used for, the discharge of water into Lake Baikal would be carefully treated so as to protect the quality of the lake. No expense would be spared, they promised, to construct the most advanced treatment plants in the country if not the world. Indeed the design included provisions for tertiary treatment with a final sand filter (Trofiuk and Gerasimov 1965, p. 53; *LG* 10/11/67, p. 12). Unfortunately the new treatment system put into use at Baikalsk was so new that there was no opportunity to test it under actual production conditions. It is hard to believe, but one critic complains that preinstallment testing was conducted "by simply using a model with 200 liters (!) of artificially polluted water." (*LG* 2/6/65, p. 1.) Equally unreal was the fact that the

planned treatment process was designed so that it included the use of a bacteria culture. Such cultures find
it very difficult if not impossible to survive in the below-
freezing weather which prevails around Lake Baikal
eight months of the year (*LG* 1/29/66, p. 2). These cultures will also die if there is any disruption in the production process due to erratic deliveries of timber supplies.
As we have seen, this has turned out to be a common
occurrence (*Kom Prav* 8/11/70, p. 2; *LG* 10/11/67, p. 12).

Not only was the treatment not thoroughly tested, but
the ministry went ahead with construction of the factory
and the opening of production lines before all the plans
had been finally approved by outside authorities. This
was in direct violation of the law of 1960 designed to
protect Lake Baikal. Moreover when approval was
obtained, occasionally it came about in something less
than an honorable manner. Thus, the Chief Sanitary
Inspector of the Ministry of Land Reclamation and
Water Management refused to certify that the treatment
equipment of the plant would protect the lake. Since
some such certification was necessary under the law
before the plant was technically allowed to open, the
Ministry of Timber, Paper, and Woodworking shopped
around until they found another inspector who was less
squeamish about signing such a certificate. When this
approval in turn was annulled, the Ministry found yet
another way to circumvent the law (*Kom Prav*, 8/11/70,
p. 2). Finally Baikalsk started operating in late 1966 or
early 1967 before treatment facilities of equal capacity
had been opened and before the staff of the treatment

plant had been properly trained (*LG* 2/6/65, p. 1; *LG* 4/10/65, p. 2; *LG* 4/15/65, p. 2; *LG* 10/11/67, p. 12; *LG* 1/17/68, p. 10).

Undoubtedly part of the reason for the haste in rushing ahead with the completion of the plants was the ministry's determination to finish construction before the whole project was cancelled. In the case of Lake Baikal, it is tragic, not trite, that "haste made waste," especially since the waste adversely affected the water. Without time for adequate testing, over 14 serious flaws were discovered in the original equipment (*LG* 1/17/68, p. 10). In a remorseful critique of the plant and its treatment facilities, some of the defenders of Lake Baikal note that even if the treatment plant were operating at its rated capacity, the effluent would, at best, be only 97 percent pure, and the plant would be able to process only two-thirds of the emitted waste (Astrakhanstev 1969, p. 102; *CDSP* 1/10/68, p. 12). And even though the yellowish, slightly smelly water that passes through the last stage of treatment is drinkable, what is good enough for the human system is still not good enough for Lake Baikal. Moreover if the treated water is so good, why is it not simply recycled and used instead of new water from Lake Baikal? When asked this question Nikolai Chistiakov, the First Deputy Minister of the Ministry of Pulp and Paper Industries (the successor to the Ministry of Timber, Paper, and Woodworking) explained that the mineral content of the plant effluent after final treatment was still five to six times higher than the fresh water of the lake. Therefore it was not as good for the production process as the pure lake water!

Because of shoddy planning, some of the more cynical critics have bitterly noted that an unplanned form of recycling is taking place anyway. Because of the peculiar nature of Lake Baikal, water circulation patterns in the vicinity of the Baikalsk paper plant move in a circle. Thus the water from the discharge pipe works its way around to the intake pipe which in any case is located only about 2 to 3 miles away (Trofiuk and Gerasimov 1965, p. 54; *LG* 4/15/65, p. 2). But however such incompetence is explained, the quality of the water used in the process of production has already been lowered. As a result, the Ministry of Timber, Paper, and Woodworking has had to provide for supplementary treatment and expense. Over 40 million rubles have been spent on the treatment plant which was originally projected to cost 20 million rubles, and still it does not function properly (*Kom Prav* 8/11/70, p. 4). Moreover, even though the treatment cost of the water had been guaranteed at 4 kopecks per cubic meter, the actual cost in 1970 was 10 kopecks per cubic meter or about 20,000 rubles a day (*Kom Prav* 8/11/70, p. 2). All of this has combined to reduce the economic feasibility of the plant.

So far, everything indicates that there will be damage to Lake Baikal even when the treatment operation operates at its rated capacity. But what has been happening in the meantime or when there are "temporary emergencies?" Unfortunately there is an answer to that

question. Several Soviet scientists have visited the
Baikalsk plant recently and have reported some dismal
findings. The treatment plant has had frequent break-
downs. Valentine Kostylev, the manager of the treatment
plant has been quoted as complaining that the filter of
the treatment plant has been malfunctioning for two
years (*Kom Prav* 9/19/68, p. 3). Gregory Galazii, the
pessimistic head of the Limnological Institute, reported
during one of his visits that the oxidizing machine at the
treatment plant was apparently inoperative, the plant's
piping seemed to be all blocked up, and the aerator had
broken down. In fact the whole treatment plant was in
the process of being reconstructed at a cost of 4 million
rubles. In the meantime waste from the plant has been
shunted to storage pools about 11 yards from the lake
shore. Black excrementlike slime has overflowed and
percolated from these pools into the lake (*LG* 1/29/66,
p. 2). During Galazii's visit, waste was flowing into Lake
Baikal at a rate of 34.5 liters per second. This apparently
had been going on for most of the year and at last report
had continued into 1970 (*Kom Prav* 8/11/70, p. 4). Poor
Kostylev remarked that "Galazii had come just at the
wrong time," but given the circumstances, it would have
been a coincidence if Galazii had come at the right time.

What was this black slime and what impact was it
having on the lake? The effluent consists of fatty acids,
methane, and organic sulphides. From 1967 to mid 1968,
the one plant at Baikalsk alone had dumped 383 tons of
such toxic substances into the lake (*Kom Prav* 9/19/68,
p. 3). As a result, islands of alkaline sewage had been

observed floating on or just below the surface of the lake at distances of 9 to 13 miles from the plant's outlet pipe, for as long as two months (Astrakhantsev 1969, p. 102). These islands—one of them 18 miles long and 3 miles wide—poisoned not only the water but the air around the lake as well (Shitunov 1969, p. 84). The effect of such toxins is not hard to imagine. The office of the Limnological Institute at Lake Baikal reported that animal and plant life near the Baikalsk sewage pipe had decreased by a third to a half (*Prav* 2/16/69, p. 1). As imperfect as the treatment equipment is, when it is not working properly, scientists at the Limnological Institute estimate that the lake deteriorates 30 times faster than it does when the equipment is operating properly (*LG* 2/6/65, p. 1).

In late 1970, I was in the Soviet Union and had a chance to talk with Nikolai Chistiakov, the First Deputy Minister of the Ministry of Pulp and Paper Industries. I must say that I came away impressed with the depth and sincerity of the Deputy Minister's anxiety over conditions at Lake Baikal. At his fingertips are daily reports about the quality of water being discharged from the treatment plant. However, as hard as he tried to reassure me that Lake Baikal would not suffer from the paper plants on its shores, and that the treatment plants would be able to process sewage in a way superior to that achieved by any other cellulose factories on our planet, the best he could ultimately do was insist that when the treatment plants were fully operational, they would fulfill the prescribed norms. Unfortunately the original water in Lake Baikal is purer than the authorized norms.

Because of the ineffectiveness of the treatment plant, the lake's defenders are seeking other ways to save it. For the most part they fall back on Galazii's original solution of a decade ago: reroute the effluent over the mountains to the Irkut River. The Ministry of Timber, Paper, and Woodworking continues to resist his proposal because this 42-mile pipeline would cost about $40 million. The expenditure of such a sum would probably make the whole operation unprofitable, which in fact it would be anyway if a meaningful value were attached to the precious water it consumes. Moreover, even if such a pipeline were built, it would only alleviate the problem at Baikalsk, not at Selenga. The plant on the Selenga is too far away to connect up to such a pipe. Accordingly, the solution of most critics for the handling of the Selenga effluent is to close down the entire plant (*CDSP* 1/10/68, p. 12).

Lamentably, at this point there is considerable doubt that Lake Baikal can be preserved even if the bypass to the Irkut River were built and the Selenga plant converted to something else. Only if the paper plants are dismantled and the towns and other industrial activity along the lake returned to a forestlike state is there much chance that the degradation of the lake can be halted. Even then the complexity of the ecological balance in the Lake Baikal basin is so complex that the damage may be irreversible (Astrakhantsev 1969, p. 103).

Because of the thin soil covering most of the watershed area, once trees are cut down it becomes very hard to regenerate plant growth in the mountains or dune areas. Reforestry work is further hampered by the hurricane force winds that are common to the region (*LG* 2/10/66, p. 2). Heedless of the hazards, Russian timber trusts have stripped many acres bare (Shitunov 1969, p. 87). The growth of towns and homes in the vicinity of the lake accelerated the process of erosion. The consequences have been far-reaching. Denuded of trees, the soil is washed away by both the rain and wind. Dust storms have even been recently reported in the area. Not only does this erosion increase the flow of silt into the lake, but it also reduces the ability of the region to retain moisture in the form of rain and snow. One observer reported in 1968 that already approximately one-third of Lake Baikal's basin had lost a significant portion of its natural water-regulating capacity (Shitunov 1969, p. 83). Thus over 130 streams and springs like the River Ude are dry a good portion of the year and flooded during the thaw (*LG* 2/10/66, p. 2; *LG* 2/15/66, p. 2). Even the unique fish of Lake Baikal find it impossible to live in dry river beds. For that matter, it is reported that from the Ude River in the extreme east to the Bukh Tarm in the extreme west, not one river has retained its original natural fish stock (Shitunov 1969, p. 87).

This disruption of Lake Baikal's ecological and hydro-chemical regime inevitably has affected the chemical makeup of the water flowing into the lake and the contents of the lake itself. Not only has the fish population been decimated, but there has also been a noticeable increase in the amount of sulphates, chlorides, mag-

nesium hydroxides, and nitrates carried into the water
by the ravaged streams. Very little of this is due to the
paper plants; most of it arises because of the destruction
of the tree cover and because of other forms of man-made
assault on the area. Consequently the elimination of the
paper plant could not in itself repair the damage.

The method of moving the cut timber causes further
damage. Large numbers of logs are rafted together and
floated or towed down the tributary rivers and through
Lake Baikal itself (*LG* 2/10/66, p. 2). Though this method
is cheapest as far as day-to-day costs are concerned, it is
expensive in many other ways. For one thing, about 10
percent of the logs sink, and sunken logs do two kinds
of harm: they cover up vital fish-breeding grounds, and
they absorb the lake's oxygen (*CDSP* 1/20/68, p. 12). As
of 1968, it was estimated that 1.5 million cubic meters of
timber had sunk over the preceding decade. With the
river bottoms covered up, 50 streams over 2,200 miles
long were eliminated as spawning grounds for fish. In
some river bottoms, the logs were piled 3 to 4 yards high.
The paper and lumber mills solemnly promised that they
would not raft their logs but would use the railroad
instead (*LG* 10/11/67, p. 12; *LG* 1/10/68, p. 12). However, shipment by rail or even by truck is considerably
more expensive. In addition the railroads have been very
slow to supply railroad cars. As a result, the logs continue
to be shipped by water and thereby continue to damage
the lake.

Stripping the soil of trees and the accelerated weather-
ing that has followed has also been responsible for

numerous landslides in the region (*Kom Prav* 9/19/68, p. 2). For years it was thought that such landslides would be unlikely because of the region's extracontinental climate (Troshkina and Volodicheva 1968, p. 112). With the disappearance of the tree cover, this is no longer the case. Most threatening of all, the cutting of the trees and the intrusion of machinery into the wooded areas has unstabilized the dunes area. By 1963, over 1,500 centners of shifting or poorly anchored sand had been noted (Goldman 1967, p. 159). Several observers report that these shifting sands have also linked up with the Gobi Desert, which is just over the border in Mongolia. Signs of accelerated movement by the desert have caused fears that the Gobi Desert will soon sweep into Siberia and destroy not only Lake Baikal, but a large portion of the taiga (*SL* 8/66, p. 44; *LG* 2/6/65, p. 1; *LG* 2/10/66, p. 2).

Conclusion

Unfortunately no one knows how to restore virginity. The only hope is that Soviet industry can curb its passion for production in order to prevent the complete degradation of the area. Toward that end, a new law regulating the use of Lake Baikal was passed by the Council of Ministers of the USSR in February 1969 (*Iz* 2/8/69, p. 2). The new law provides for the establishment of 20,860 square miles of a special water conservation zone where no timber may be cut (*SL* 8/70, p. 50). In addition no timber is to be cut on mountain slopes steeper than 25 degrees. Submerged logs are to be removed from the rivers and Lake Baikal. Tractors are no longer permitted to drag logs down the mountainside; cableways are to be used instead. Construction of new factories is to be strictly regulated, and none are to be located where they might pollute the lake. In addition, the operators of the Baikalsk cellulose and paper plant were ordered to complete the construction of their purification plant before the end of 1969 (*Kom Prav* 2/11/69, p. 1). The Selenga Cellulose carton plant was ordered not to start production until its purification equipment was ready and similarly restricted.

The question that remains, however, is whether this law was soon enough or strong enough. The first thing to remember is that the 1969 law is not the first dealing with Lake Baikal. A glance at the 1960 law shows that it addressed itself to many of the same issues that are covered in the 1969 law. From what we have seen, the first law has proven to be ineffective, and there is not much reason to believe that the 1969 law will ultimately be any more so. The cellulose and paper factories were not supposed to operate without adequate water treatment; trees were not supposed to be cut on mountain slopes near the lakes, etc. Will things be any different now? Will the Ministry of Timber, Paper, and Woodworking sit by obediently with an idle production line when the Selenga plant is ready to operate if the treatment plant is still not fully operational? If so, they have greater self control than their colleagues at Baikalsk. In fact, press reports in *Komsomolskaia Pravda* of August 1970 indicated that abuse of the lake had increased, not decreased, since the passage of the 1969 law (*Kom Prav* 8/11/70, p. 2).

Nor can it be assumed that worse will not follow. While Lake Baikal may no longer be a virgin, it still is much purer than anything else around. It is only natural, therefore, that various proposals continue to be advanced by Russian industrial suitors. For example, one proposal—fortunately not acted upon—was that Lake Baikal be breeched in order to accelerate the accumulation of water in the Bratsk reservoir. This would have resulted in a drop in the level of Lake Baikal with unknown consequences (Buiantuev 1960, p. 9). More of a threat is the hint that Selenga and Baikalsk are not the only cellulose and paper mills that the Ministry of Timber, Paper, and Woodworking would like to build in the lake basin. At least one other plant is apparently in the advanced blueprint stage (Buiantuev 1960, p. 14; *LG* 10/11/67, p. 12). Similarly the chemical ministries have let it be known that the Soviet economy would benefit greatly if chemical plants could be established on the lake's shores and could utilize the lake's waters. What's good enough for the cellulose and paper industry should be good enough for the chemical industry. God forbid that oil should be discovered nearby. If industrial officials have their way, there is no disputing Galazii's assertion that Lake Baikal's singularity will be destroyed within the next decade.

The *story* of Lake Baikal is not unique—similar tragedies take place all over the world. But Lake Baikal *itself* is, or was unique, and somehow or other it should have been spared. One would have thought that Lake Baikal would have been safe in the USSR. This is not only because the USSR is a vast country and the lake is in a remote region, but because in Soviet society, all the country's natural resources and means of production are owned by the state. Unfortunately the rape of Lake Baikal shows that public greed and lust can be as destructive as private greed and lust.

Postscript
On September 24, 1971, after the manuscript for this book had gone to press, yet a third decree was issued about Lake Baikal. This time it was signed not only by the Council of Ministers of the USSR but by the Central Committee of the Soviet Communist Party as well. Presumably the involvement of the Communist Party indicates *real* concern. Instead of pious statements and promises about the care being shown by industry in the Lake Baikal Basin, the new decree implicitly acknowledges that conditions are in a deplorable state and that the previous two laws were largely ignored or at best, had little impact. For instance among other provisions the new law, demands that—

1. The Baikalsk sewage treatment plant must be fully operational and meet specified norms sometime in 1971. It was supposed to have been ready in 1969 according to the law of 1969 and industrial officials have always given the impression that whatever was wrong was an aberration from the normal effective operation.

2. The sewage plant at the Selenga Cellulose Carton

Combine must be fully completed in 1972, before the first production operations are due to begin.

3. All enterprises in Lake Baikal Basin must draw up a timetable for the construction of sewage treatment facilities. All facilities should be operational no later than 1972 except those in the city of Ulan Ude, which has until 1973.

4. Timber cutting and shipping procedures must be improved so that erosion will be reduced, and the sinking of logs to the river and lake bottom will be halted. Specific plans for timber cutting for 10 years in the future must be drawn up to ensure that timber is selected with the least impact on the environment.

5. During 1971–1975, sunken timber in the beds of several major river tributaries and in the bottom of the lake must be removed.

6. Other miscellaneous steps must be taken to ensure that the high quality of the water in the basin is maintained and that all these orders are complied with.

Unfortunately there is abundant reason to fear that even the direct intervention of the Central Committee of the Communist Party has come too late to prevent further damage or bring about the restoration of Lake Baikal. Doubtlessly, conditions will not deteriorate as fast as it appeared would be the case in June 1971. For that the world is grateful, but we have seen unfulfilled promises and unimplemented laws before, and Lake Baikal is too fragile to withstand even the mildest maltreatment.

7 Ecological Facelifting, or Improving Nature

Introduction

From earliest recorded time, not only has man struggled against nature, he has sought to reshape it to suit his purposes. The magnitude of his surgery has been limited only by the extent of his technical competence. This competence has not increased in linear fashion. Frequently earlier generations were capable of efforts beyond the reach of their descendants. But whatever the engineering or planning skills of a particular generation, the intent was generally to make life more comfortable or more profitable. All too often, however, difficulties would arise when one man's definition of comfort and profit differed from another's. Sometimes such differences would develop between contemporaries and sometimes between sons and fathers or grandfathers. Moreover the attempt to improve on nature in one sector more often than not set in motion other forces which sometimes have brought an even greater need for remedial action than there was in the first place.

   Although we are primarily concerned here with how the Soviets have sought to rearrange nature, it is readily conceded that such schemes are not a monopoly of the communist countries. For example, we in the United States almost succeeded in severing Florida from the mainland with a canal. The so-called North American Water and Power Alliance had a grandiose plan to reroute numerous rivers from the northern part of our hemisphere. And without any plan, the United States managed to create a dustbowl out of its midwest by overcultivation and a dying lake out of Lake Erie by overindustrialization. Redoing nature, intentionally or unintentionally, is an age-old process and it generally plays an important part in the process of economic development. Unfortunately, after an initial project is completed, a subsequent reordering of nature is often felt to be necessary in order to compensate for the destruction the planners and builders generated with their first project.

No matter how simple they may appear to be at first, man's efforts to improve on nature often turn out to be complex undertakings with unexpected ramifications. It is like tampering with the free market. When you make an adjustment in one sector, somehow or other there are aftereffects in another sector. Friederich Engels saw the essence of the matter with remarkable clarity in the nineteenth century. (A longer selection from his writings appeared in Chapter 1.)

"Let us not, however, be very hopeful about our human conquest over nature. For each such victory, nature manages to take her revenge. Each of these victories, it is true, has a first order of consequences which we can anticipate. But in the second and third orders (secondary and tertiary effects) there are quite different, unforeseen effects which only too often cancel out the significance of the first." (Engels 1940, pp. 291–292; Engels 1955, pp. 140–141.)

Engels' statement has come to have particular meaning today. First of all, some of the same incidents described by Engels in the 1880s are reoccurring in the world, including, as we shall see, the USSR. Second, as serious as the secondary and tertiary effects were several decades ago, with our vastly more powerful technology the potential for damage is even greater now. When we consider Engels' concern over this question, it is ironic that the highest potential for increased destructiveness comes when the state monopolizes all the productive powers of the country into a single decision maker's hands, as often happens in a communist country. To some extent the drive to redo nature is no different in communist countries from what it is in noncommunist countries, especially those that are underdeveloped. In a communist country, however, the opposition or resistance to the restructuring power of the state is likely to be considerably less because of the absence of private property or opposition parties. With such concentrated power at its command, the potential a communist state has for reordering nature is unprecedented.

Finally as important as the economic, political, and corrective factors may be for explaining why man attacks nature in such a vigorous way, there may also be a psychological element, especially in the case of Russia. Citizens in large land masses as well as planners in developing countries generally seem to have an immense fascination for large water bodies. In the first instance this may be an attempt to make up for the absence of large ocean bodies which most other countries of the world are able to enjoy; in the second instance, it may just be that unharnessed water power constitutes a mocking reminder of a country's impotence. Alternatively, as Peter Wiles has put it, this excessive concern for water may be nothing more than a Freudian response. Whatever the precise explanation, the interplay of these economic, political, and psychological forces is especially well illustrated by a study of what the Russians have been doing to the Caspian and Aral Seas.

In the Beginning

The Caspian and Aral Seas are unusual water bodies. Unlike most seas or lakes in the world, the flow of water is one way—in. Water is not carried out of either of these major seas to other rivers or seas or oceans. Both water bodies are located in very dry regions of the world where the evaporation rate is exceptionally high. The rapid evaporation absorbs the incoming water, which explains why the seas do not overflow, even though there is no outlet. In the case of the Caspian Sea, the rapid evaporation rate also makes possible a valuable mineral recovery operation. Water is drawn off the Caspian into the Gulf of Kara-Bogaz-Gol where evaporation is especially rapid, and the residue is converted into valuable minerals and salts. The general characteristics of both seas are presented in Table 7.1.

Even more intriguing, both the Caspian and Aral Seas have had rather erratic geological lives. Researchers believe that the Aral Sea was dry only 30,000 years ago. Subsequently a sea was formed and then several centuries ago, around the twelfth century, the Aral Sea contracted, only to expand again during the thirteenth and fifteenth centuries (Shnitnikov 1968, p. 168). In somewhat the same way, the water level of the Caspian has changed

markedly over the last few centuries. In the second
century B.C., the shoreline was 32 meters below sea level
(Shnitnikov 1968, p. 76). Villages were built along the
edge of the sea only to be submerged as the Caspian rose
in the centuries that followed. By A.D. 1400, the sea depth
increased to 22 meters below sea level which, it is
believed, was about the depth that existed in 2000 B.C.
(Gumilev 1964, p. 60; Shnitnikov 1968, p. 76.) Except for
a slight dip in the sixteenth century, the level of the
Caspian was fairly stable until the mid nineteenth century (Dobrovol'skii et al. 1969, p. 135).

The changes in the level of the Caspian and Aral Seas
up until the mid nineteenth century are explained primarily by variation in rainfall and temperature in the
watersheds of both seas (Apollov 1962, p. 80; *Kom Prav*
11/22/68, p. 2). In 1929 the level of the Caspian Sea and
in 1960 the level of the Aral Sea began to fall rapidly at
rates that were clearly due to more than just changes in

Table 7.1 Characteristics of the Caspian and Aral Seas

	Caspian Sea	Aral Sea
Elevation	−28 meters	53 meters
Area		
low water	372,000 square kilometers	64,000–66,458 square kilometers
high water	424,000 square kilometers	
maximum depth	1,020 meters	68 meters
average depth		16.4 meters
volume		1,023 cubic kilometers

Sources: *SGRT* May and June, No. 5–6, 1965, pp. 325, 328; Shul'ts 1968,
p. 489; Malinkevich and Belianov 1966, p. 117.

rainfall (Apollov and Bobrov 1963, p. 72). From 1847 to 1928 the level of the Caspian fell by 56 centimeters, or about 6 millimeters a year; from 1929 to 1965 it fell by 2.47 meters (8.2 feet), or 67 millimeters a year (Dobrovol'skii et al. 1969, p. 128). By 1970 it had fallen another 0.1 meter so that the total drop was 2.6 meters ($8\frac{1}{3}$ feet) (Iordanskiy 1970, p. 37).

The recent history of the Aral Sea is similar. From 1960 to 1967 the water level fell 1.76 meters (5.5 feet), or about 24 millimeters a year (*SGRT* 1969, No. 3, p. 146). This rate is expected to continue, so that by 1980 the sea will have fallen a total of 4 to 4.5 meters (*Prav* 11/7/68, p. 2). Since the average depth of the Aral Sea is 16.4 meters (50 feet) and its maximum depth is only 68 meters, it is estimated that by the year 2000 the Aral Sea will have turned into a salt marsh (Shul'ts 1968, p. 489; Kosarev 1970, p. 13). Even by 1980 it is expected that the area of the sea will have been reduced from the 64,000–66,000 square kilometers it was in 1960 to 15,000–25,000 square kilometers, and its volume will have been reduced from 1,023 cubic kilometers to 180–190 cubic kilometers (Shul'ts 1968, p. 490).[1]

Economic Development and Environmental Disruption

Whatever natural explanation there may have been for fluctuations in the level of the Caspian and Aral Seas, by early 1930 in the case of the Caspian and 1960 in the

[1] The conversion of the metric scale into the linear system is sometimes rounded off since so many of the original figures are approximate sums anyway. See Conversion Table, Appendix D, p. 331.

case of the Aral, manmade factors became more important. As shown in Table 7.2 (see Dobrovol'skii et al. 1969,
p. 132) the area of the Caspian Sea was fairly steady
from the years 1917 to 1933. Except for a brief respite
during World War II, from 1933 on the Caspian fell
noticeably. This coincides with the spread of industrialization and urbanization and the attempt to increase agricultural output.

The role of man as the prime mover becomes apparent
when we examine what happened to the sources that
supplied the sea. Just as a pinched air hose can cause a
deep-sea diver to suffocate, so a blocked or diverted river
can cause a sea to shrink. Evaporation in Central Asia
is intense and in the absence of cool weather or unusually
heavy precipitation, any diversion of water has a marked
effect. This is an especially acute situation in Central
Asia, since both the Caspian and Aral Seas are heavily
dependent on only a few rivers. For example the Aral
Sea is fed by only the Syr Darya and the Amu Darya
Rivers which originate in the mountain ranges that
separate the USSR from Afghanistan and China. A
larger number of rivers flow into the Caspian, but except
for the Volga and its tributaries, none of them amounts
to much. Rivers provide 80 per cent of all the water
added to the Caspian each year and the Volga makes up
80 percent of the river flow (Bobrov 1961, p. 49; Dobrovol'skii et al. 1969, p. 133). So just the diversion of the
Volga, Syr Darya, and Amu Darya could and did have
disastrous consequences for the Caspian and Aral Seas.
Why did these diversions take place?

Table 7.2 Area of the Caspian Sea (square kilometers)

Year	Area	Year	Area
1906	403,250	1936	394,340
1907	402,680	1937	392,360
1908	402,680	1938	387,710
1909	403,250	1939	382,940
1910	401,640	1940	379,440
1911	399,470	1941	377,950
1912	399,560	1942	380,230
1913	398,570	1943	380,710
1914	399,380	1944	380,230
1915	401,180	1945	377,370
1916	402,000	1946	378,480
1917	402,460	1947	380,080
1918	401,000	1948	381,030
1919	400,820	1949	379,760
1920	399,650	1950	376,900
1921	398,390	1951	374,760
1922	397,490	1952	374,570
1923	396,950	1953	373,240
1924	396,950	1954	372,660
1925	395,960	1955	371,710
1926	396,770	1956	370,940
1927	398,480	1957	372,470
1928	400,100	1958	375,150
1929	401,640	1959	375,910
1930	400,550	1960	374,760
1931	399,380	1961	371,520
1932	400,190	1962	369,700
1933	399,920	1963	371,520
1934	397,940	1964	372,850
1935	396,230	1965	372,100

Source: Dobrovol'skii et al. 1969, p. 132.

Even though capital can be put to many alternative uses in a developing country, all too often the project with a high and immediate payback takes a back seat to projects with a longer payback period. This is particularly so if the slow return project also provides the planners with a chance to reshape their country's water courses. In fairness, it must be acknowledged that massive water projects are a means of using readily available quantities of unskilled labor in order to build up a country's infrastructure. But as mentioned earlier, of equal if not greater importance in commissioning water projects is the primeval or mystical urge that seems to be brought out in men by unused streams and water bodies. Although planners in almost all countries are affected with this peculiar passion for power planning, the reaction is perhaps strongest in the developing countries. To a poor country, idle rivers serve to remind it of its backwardness. Some planners may view the river as a source of fertility and power. Others may simply conclude that since other countries have developed that way, they too must dam up their own rivers.

In addition to whatever else he may be known for, Stalin will go down in history as the greatest rearranger of men and nature the world had ever seen, at least as of 1953. Other rulers have had dictatorial power to give orders and divert manpower and equipment, but until Stalin no one had the technology as well as the power to carry out such huge undertakings over such a vast expanse. Whether or not such action was economically warranted was inconsequential to Stalin. With Stalin's severe case of what we can call aquaphilia, the results were inevitable.

Once it was decided to redo the rivers of the Soviet Union, it was natural that the Volga would be singled out for special attention. Not only did the Volga slice through the center of the country, but it was the longest river flowing south in the country and the longest Soviet river west of the Urals. Naturally Stalin saw the Volga as the centerpiece in his overall design. In the early 1930s, it was announced that 13 big dams would be built along the Volga. These dams and the reservoirs that were to form behind them would make possible a variety of projects. The most obvious were electrical power and water for irrigation. In addition canals were to be dug and river channels enlarged. By March 1937 Stalin had managed to link Moscow to the Volga and the Caspian Sea. With the connections between the Volga and the White and Baltic Seas which had been completed in June 1933, this meant that Moscow was also joined by water with the north. When Stalin completed a canal linking the Volga and the Don River in July 1952, ships from Moscow could also sail to the Azov and Black Seas. In this way, water from the Volga not only transformed Moscow into a port of five seas, but it provided Moscow and Central Russia with direct access to an increased water supply for industry and municipal needs (Map 7.1.)

While these projects provided some important benefits to the country, they also brought with them serious shortcomings. In addition to the often valuable land lost behind the newly erected dams and reservoirs, a good

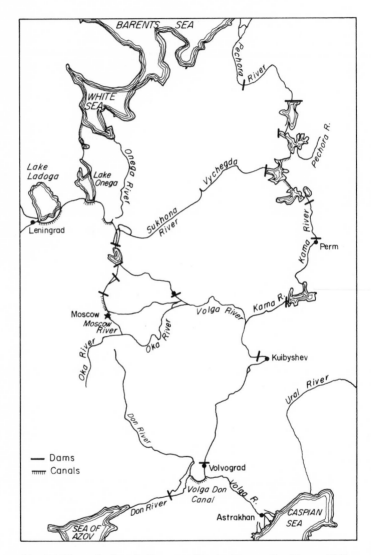

Map 7.1 The main rivers in European Russia. (Source: Sarukhanov 1961, p. 54)

deal of water was lost. Some of it disappeared into underground channels; but large quantities simply evaporated into the air. When combined with the increasingly large quantities of water diverted from the rivers to the irrigation canals, the river flows had diminished sharply by the time they reached the sea.

Diversion of the Aral's arteries began after World War II with the siphoning off of water from the Syr Darya for irrigation in the Fergana Valley, an oasis in Uzbekistan, Tadzhikstan, and Kirgizia (Shul'ts 1968, p. 489). By 1965, almost all the water from the Syr Darya had been tapped for irrigation so that it was virtually dry by the time it reached the Aral (*Kom Prav* 9/16/65, p. 2; *Kom Prav* 4/7/70, p. 6). Only the Amu Darya River had any water left to feed the Aral, and it too was shrinking rapidly (Map 7.2). In 1962 a canal was built from the Amu Darya to the Karakul' Oasis (*SGRT* 1970, No. 6, p. 511). By 1968, 37 canals were already feeding on the Amu Darya, and plans had been announced for the construction of 18 more in the years to come (*Kom Prav* 8/11/68, p. 2). For example, in 1969 construction began on a major dam at Tyuyamuyan on the Amu Darya in order to provide more water for irrigation. In 1970 additional work was performed on the Amu-Bukhara Canal for the same purpose (*SGRT* 1970, No. 1, p. 60; *SGRT* 1970, No. 6, p. 511). With the main sources of their supply diverted, the Aral and Caspian naturally suffered.

To many Soviet economists, much of what was done was regarded as needless waste. Although economists are not immune from the passion for power planning, it is

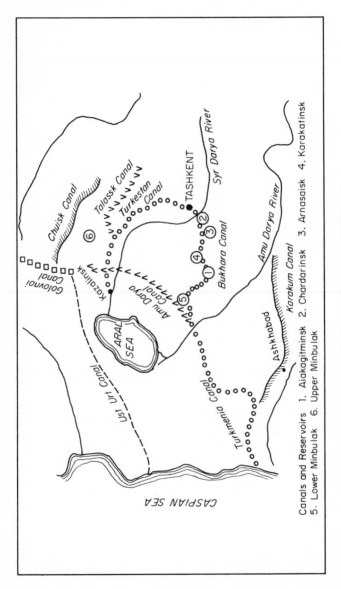

Map 7.2 Rivers and canal network in Central Asia. (Source: *Kras Zvez* 1/17/71, p. 4)

possible that if a system of more meaningful prices and economic valuations had been in use as they suggested, the ecological damage would have been less. The fact that no payment had to be made for the land that was flooded made it easier for the engineers to engulf large areas of formerly valuable land. Several Russian economists estimate that in Central Asia, as much land has been lost through flooding and salination as has been added through irrigation and drainage (Armand 1966, pp. 40, 81; Gerasimov 1968, p. 448). Presumably more efficient use of the land would have increased productivity enough to offset the need to invest so heavily in irrigation and flooding. Despite these criticisms of the economists, Soviet engineers envision irrigating another 8.3 million hectares in addition to the 4.6 million hectares already under irrigation in 1967 (Kes' 1967, p. 87).

Similar questions were also raised about the value of making Moscow into a port of five seas. The Volga-Don Canal reportedly has not been heavily used and apparently will never justify its cost. Again, if more appropriate cost estimates had prevailed and if Stalin had allowed someone to pay attention to such matters, there is the slimmest chance the canal would not have been built, thereby lessening the drain on the Volga. (Realistically, however, as long as Stalin was in charge, it is likely that no cost system would have made much difference.)

A Cause for Concern
Whether or not a fall in the Caspian and the Aral Seas could have been averted or alleviated, the fact remains

that the water levels of the two seas fell. When it became clear what was happening, many scientists expressed alarm. Why were they so concerned? (Apollov et al. 1963, p. 10.)

Any sudden change in one's environment is unsettling. Not only is it psychologically disturbing, it may also be physically inconvenient and financially costly. Usually man is able to adapt himself gradually to new sets of conditions. Should these conditions change again, then a new process of adjustment must take place. However, when the nature of the change is not just seasonal but long run, with no indication of where it may end, the effect can be particularly unstabilizing.

The reduction in the flow of water to the Caspian and Aral Seas had an immediate effect and was quickly noticed. The reason for this is that a portion of the Caspian and all of the Aral Sea are very shallow. Consequently any reduction of water immediately shows up as exposed land. By 1966 the area of the Aral Sea had diminished by 10 per cent from what it had been in 1960 and was shrinking rapidly. Since the southern part of the Caspian had depths of up to 3,280 feet, contraction in this area is not much of an issue. However, as shown in Map 7.3, it is a substantial matter in the northwest, north, and northeast (Apollov 1962, p. 81). Frequently maximum depths of only 15–30 feet are reported and it is here that the sea has retreated the farthest. Every time the level of the sea drops by one meter, there is only a reduction of 0.5 percent in the total surface area in the south, whereas in the north as much as 17 percent

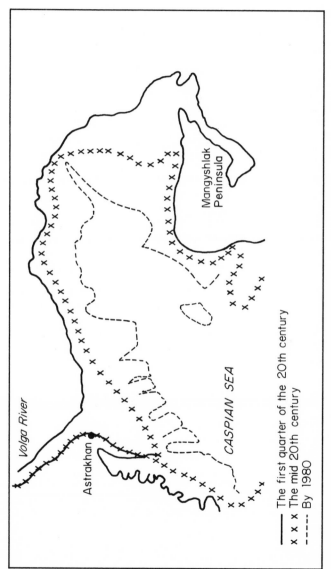

Map 7.3 Shrinkage of the Caspian Sea. (Source: Apollov 1962, p. 81)

of the surface disappears (Bobrov 1961, p. 49). The retreating sea has also affected mineral operations in the Gulf of Kara-Bogaz-Gol (see Map 8.1 in the next chapter). For years as the sea water evaporated, the shallow gulf had been a major source of valuable minerals and salts such as sulfur nitrate, bromides, chlorides, and sulfur magnesium. But as the Caspian has fallen, the channel into the Gulf has shriveled up. The flow of water into the Gulf has been reduced from 20 to 9–9.5 cubic kilometers a year. As a result the maximum depth of the Gulf has fallen from 13 to 3 meters (Stas' 1968, pp. 108–109; Buinevich 1969, p. 33.) Moreover some authorities who are concerned with the Caspian argue that the flow of water into the Gulf should be further reduced or eliminated completely to prevent needless evaporation and loss of water from the sea (Vendrov et al. 1964, p. 32). They are opposed by those whose main interest focuses on the Gulf and who argue that a valuable source of minerals is being lost (Buinevich 1969, p. 32).

What else has happened as a result of the drop in water levels? The damages include pollution of the seas, a drop in the fish catch, an increase in disease, loss of seaports, and reduction in farm land.

One way to reduce or eliminate some types of water pollution is to dilute the effluent until it reaches nontoxic levels. Naturally if the water flow is reduced, there is much less dilution so that the dissolved oxygen in the water falls rapidly. In the same way the relative concentration of pollutants rises. The problem of water pollution is particularly serious in the Caspian Basin where there is more industrial activity than in the area around the Aral. It is probable that even if the flow of water into the Caspian had not been reduced, it would still have become polluted. The pollutants originate from emissions on the sea itself and from its tributaries.

The Caspian is a major source of oil. An entire city has been constructed several miles offshore and numerous oil wells have been drilled beneath the sea. Nowhere in the world is this done without oil spills, and the Caspian has not been spared blowouts and leaks at its wells. At one time an oil spill stretching 2,000 square miles was reported on the Caspian (Simonov 1970, p. 59). In January 1971 a major fire raged out of control for almost a month at one of the rigs. After the fire was extinguished debris and oil slicks covered vast areas of the sea (*Prav* 1/2/71, p. 3; *Sot In* 2/6/71, p. 4). Oil discharges also emanate from several of the refineries along the banks of the Caspian. For years virtually no effort was made to treat the waste of the refineries. It was reported in 1966 that 47,000 metric tons of oil were dumped into the Caspian from the Republic of Azerbaijan itself (*Priroda* 1/68, p. 115). Nor is the natural state of the sea improved by the blasting and seismic explosions that have been used since 1941 in the search for more oil (*LG* 9/27/66, p. 2).

Conditions deteriorated so much that in October 1968 a special resolution of the Council of Ministers was adopted in order to prevent further pollution. Entitled "On Measures to Avert the Pollution of the Caspian

Sea," the resolution sought to accelerate introduction of preventive measures and the construction of treatment plants at 100 refineries and factories along the sea, as well as at 14 cities. All of these installations had been dumping their waste into the water for years without any restraint (*Prav* 10/3/68, p. 2). The new law also sought to stop oil tankers from discharging their ballast into the sea (*Bak Rab* 6/12/68, p. 2). Special discharge and treatment receptacles were built at selected cities along the coast in order to provide some place other than the sea for the tankers to dump their ballast (*Prav* 10/3/68, p. 2; Simonov 1970, p. 59). While there has been some improvement, there are indications that as yet all these efforts have not been completely successful and that oil and sewage discharge into the Caspian continues to be a hazard (Iordanskiy 1970, p. 41). As recently as March 1971, there were reports that major cities such as Baku, Makhachkal, Krasnovodsk, and Gousan were still dumping untreated raw sewage into the sea (*Prav* 10/3/68, p. 2; *Prav* 9/20/69, p. 2; *Turk Isk* 9/20/70, p. 2; *LG* 3/3/71, p. 11).

If and when the emission of oil and other pollutants along the sea is brought under control, the battle will have only begun. Oil also flows into the sea from several of the major rivers that feed the sea. For example, in March 1971 *Pravda* reported that an oil pipeline broke along the Ural River (*Prav* 3/20/71, p. 2). This pipeline carries oil from the Mangyshlak Peninsula alongside the Caspian to oil refineries near Kuibyshev along the Volga River. *Pravda* reported that barriers had been erected to contain the oil but that oil was nonetheless seeping into the river, which in turn would carry it down to the Caspian.

Until this incident the main flow of oil into the Caspian came from the Volga. For years only feeble efforts were exerted to force factories and municipalities to treat their waste. Since the banks of the Volga are heavily industrialized, this meant that a variety of industrial effluents, particularly oil, were sent off in enormous quantities to the Caspian. In 1966, factories and cities along the Volga discharged over 300,000 cubic meters of untreated sewage into the river each hour (Kazantsev 1967, p. 36). One authority estimates that the Volga carries almost one-half of all the discharged effluent in the USSR (*LG* 3/3/71, p. 11). Particularly bothersome were the several refineries which either had no or frequently inoperative facilities (*Turk Isk* 3/3/71, p. 3). The concentration of oil has reportedly been so great that on one occasion at least, the Volga caught fire. As one newspaper put it, "The children set the Volga on fire and the firemen came and put out the river!" (*Sot In* 7/4/70, p. 2.) Volga steamers now carry signs forbidding passengers to toss cigarettes and lighted matches overboard (*Krokodil* 8/70, No. 24, p. 4). One writer explained such signs as evidence of the concern about litter prevention; but given the river's incendiary past, it is more likely a sign of fire prevention.

The reduced flow of water and an increased discharge of oil and other industrial and municipal wastes combined to create an environment that can hardly be considered

hospitable for water life. Fish kills along the Volga and its
tributaries are a common calamity. The Volga and the
Ural (the second largest supplier of the Caspian) are
both periodically classified as dead rivers (*Bak Rab*
6/12/68, p. 2). Today because of the pollution and dam
construction along the Ural, it no longer carries fish to
the Caspian (*Prav* 7/12/67, p. 3). Inevitably this affects
fish life in the Caspian itself. That this entails a significant
economic cost becomes apparent when it is realized that
prior to 1929 and the fall of the water level, the Caspian
yielded up to 40 percent of the fish catch in the USSR
(Bobrov 1961, p. 47). Furthermore the cost of catching
fish in the Caspian was 40 to 50 percent of what it was in
the Pacific (Apollov et al. 1963, p. 11).

The decline in the fish haul in the mid 1930s can be
seen in Table 7.3. In 1967, the total fish catch was less
than one-half of what it was in 1936. The figures are
drawn from different sources, but the sturgeon catch by
1970 was about one-quarter of what it was before the
revolution. In some areas fishing for sturgeon has been

Table 7.3 Fish Catch in the Caspian (in 1,000 metric tons)

Year	Total	Sturgeon	Sudak	Sprats
1900		30–40		
1936	500	21.5	55.2	5.3
1950	332	13.5	31.4	21.6
1960	387	10.1	14.6	176.0
1967	230			

Sources: Apollov et al. 1963, p. 130; Dobrovol'skii et al. 1969, p. 254; *EG*
3/68, No. 9, p. 19.

banned (*LG* 3/3/71, p. 11). Because of the shortage, the price of caviar has risen so high that one Soviet paper complains that some Soviet citizens have taken to caviar rustling. The poachers sneak out illegally in speedy motor boats and sell their catches at speculative prices (*Zar Vos* 2/2/70, p. 2.)

The fall in the catch of the more valuable fish has been partially offset by an increased haul of cheaper fish such as sprat. Still intensive efforts have been made to restock and regenerate the more valuable fish stocks. The poorer fish species, however, are often an important link in the food chain of the richer species. Consequently the increased depletion of the less desirable fish is likely to handicap even more the recovery and return of fish like the sturgeon, which only thrive when there is an abundant base for them to feed upon. This is an important question. The sturgeon, after all, is the source of caviar and caviar has provided a valuable supply of foreign exchange. Since the Caspian is the source of 80 to 90 percent of all the world's sturgeon and 90 to 95 percent of the world's black caviar, the maintenance of the Caspian as a fertile ground for the sturgeon is a priority matter (*Bak Rab* 6/12/68, p. 2; Bobrov 1961, p. 47).

The catch of fish has been affected not only by a deterioration in the quality of the water but by a disruption of the fish breeding habits. The erection of dams has severed the migratory and spawning habits of the fish that breed in the freshwater streams and then move into the seas. Some dams on both the Volga and Ural have been equipped with fish gates, but there is considerable

debate about their effectiveness. Moreover, the dams
that were designed to prevent the annual spring floods
simultaneously prevent the nourishing of traditional
breeding grounds in the deltas of the various rivers. Thus
new sites have had to be found as some of the old sites
have been left high and sometimes dry or swampy. Even
the new lower-lying breeding areas are sometimes affected
by inadequate supplies of water. This happens when the
dam operators violate their operating instructions and
release less water than stipulated. The dam engineers
are likely to do this whenever the spring flow is somewhat
less than normal. (*EG* 3/68, No. 9, p. 19; Kolbasov 1965,
pp. 145–149; *LG* 6/17/70, p. 11; *CDSP* 10/12/66, p. 15.)
The overall effect of the curtailed flow of fresh(?) water
into the Caspian from the Volga and Ural has also caused
the saline content of the water to increase to what the
Russians describe as 13 percent. (It is not clear what the
13 percent is a percentage of.) In some areas, levels as
high as 20 percent have been reported. (For a contrary
view, see Dobrovol'skii et al. 1969, pp. 185–188.) The
impact on fish life can be deduced when it is realized
that the same authorities report that a saline content of
4 to 10 percent is appropriate for fish while 4 to 7 percent
is considered to be the optimum. It is not only the lack of
water that affects aquatic life in the Caspian. Just like the
Aswan Dam in Egypt, dams on the Volga also hold back
the flow of silt and nutrient which is an essential source of
food for the water life population in the Caspian. (*EG*
3/68, No. 9, p. 19; *EG* 7/66, No. 30, p. 21.)

Although the quality of the fish native to the Aral Sea

was not as high as it was in the Caspian, the impact of
polluting and shrinking the Aral has been even greater.
From a typical haul of 40 metric tons in 1962, the catch
dropped to 20 metric tons in 1967 (*SGRT* 1969, No. 3,
p. 146). Apparently by 1970 it had fallen to 6–8 metric
tons or to 20 percent of what it had been (*Sot In*
8/15/70, p. 2). And as the salt content of the sea rises,
the expectation is that the remaining fish in the Aral will
rapidly be annihilated (*Kaz Prav* 2/6/69, p. 2).

This disruption of the environment can also have other
unexpected and undesirable effects on the region's
ecology. Fish breeding grounds were adversely
affected and swamp conditions were created along the
Karakum Canal and the Amu Darya when water was
drawn off for irrigation and contaminated with pollutants. This eliminated most of the fish but not the mosquito larvae which some of the fish used to consume. As
a consequence there was a sharp rise in the mosquito
population, including some that transmitted malaria.
As a result the disease has returned to Ashkhabad after
having been gone for many years (*Turk Isk* 9/16/69,
p. 3). In an effort to solve the problem it was proposed to
introduce a fish called the Belyi Amur which feeds on
mosquito larvae. The drop in the quantity and quality of
water in the Volga is also an important factor in explaining why Astrakhan, a major city at the mouth of the
Volga, was one of the first cities in the USSR to report
the outbreak of cholera during the epidemic of 1970
(*NYT* 8/7/70, p. 10; *Sot In* 8/22/70, p. 3).

The deterioration of the Caspian and Aral Seas has had

other less severe consequences as well. Several fishing villages once located on the shores of the Caspian now find themselves located as far as 30 miles from the shore (Bobrov 1961, p. 50). Onetime steamship ports such as Ilychik and Astara have had the same experience. Access to Muinak and Aral'sk in the Aral Sea has also been blocked (*SGRT* 1969, No. 3, p. 146). This has necessitated extra expense for dredging and in some cases the relocation of the ports (Bobrov 1961, p. 51). There is now only one navigable channel in the Volga delta (Apollov and Bobrov 1963, p. 69).

The falling Caspian has also created turmoil for the oil drillers. They must decide whether to locate close to the water on the assumption that the level of the sea will fall another meter or two (three to six feet) by 1980 and perhaps by a comparable amount by the year 2000. Alternatively, some might decide to locate a respectable distance above the water in case the sea should return to its previous heights (Kosarev 1970, p. 13; *Prav* 10/7/68, p. 3).

The Caspian has been subjected to so many forces that it is sometimes difficult to interpret their impact. For instance, in 1965 a report appeared in the scientific journal *Priroda* (Nature) about the discovery in the Caspian of an unusually high recording of what was called natural strontium, which in itself is not dangerous (Timoschchuk 1968, p. 90). Because of ambiguities in the article, it is unclear if the reference is to strontium 90. V. I. Timoshchuk, the author of the article, implies that the recorded levels are higher than those recorded in

other seas of the world and that the strontium content
is particularly high on the eastern shore. According to the
article, strontium concentrations range from 4.24 milligrams per liter on the western shore to 17.10 milligrams
per liter on the eastern shore. In some places readings as
high as 25 milligrams per liter have been reported. According to one estimate made by officials in the Division
of Radiological Control in the Massachusetts Department
of Natural Resources, this amount indicates doses of strontium several hundred million times higher than anything
ever recorded in water bodies in this country. The only
exception would be water directly affected by atomic
testing, processing, or disposal. If such readings have
been correctly interpreted (the Massachusetts officials
find it hard to believe that such high readings are indeed
correct), this suggests that the Caspian is an unusual sea.

Less spectacular, yet of equal concern, is the uncertainty
over what effect the shrinking seas will have on the
weather. Evaporation patterns are being changed as new
water bodies are being created and old ones dissipated.
This in turn affects wind circulation and surface water
accumulation. Water bodies tend to temper the continentality of the climate. Already there are reports that
because of the contraction of the Caspian Sea, the climate
around Mangyshlak has become more continental with
longer winters and shorter summers (Apollov 1962, p.
70). Although some specialists have argued that only
the temperature around the narrow coastal strip will be
affected, the same changes are occurring around the
Aral Sea (Shul'ts 1968, p. 490). Moreover, since the

disappearance of the Aral is much more imminent, there is also concern about the impact on humans of the disappearance of this water body. In all likelihood, the area will become a desert with salt and dust storms (Shul'ts 1968, p. 490; *Kom Prav* 8/11/68, p. 2.) There is considerable fear that the existing population will find it too difficult to live in such surroundings without the tempering presence of a large body of water. As a result there may well be a large-scale migration as the water disappears.

Because of the damage the shrinkage of the Caspian and Aral was causing, it was natural that several Soviet scholars would suggest proposals to restore the health of the seas. In other words, now that man has had his negative impact, what could he do to offset this damage? If enough money is available, solutions can always be found. Nature takes its revenge, however, because invariably these new corrections tend to generate their own negative consequences which require yet additional rectifications. It is to these proposals that we now turn.

8 Reshaping
the Earth

In the literal sense, some of the cures that have been proposed for the Caspian and Aral Seas have earth-shaking implications. The "solutions" range from doing nothing to altering the makeup of the rivers and seas in Siberia. Some of the plans are new and some are 100 years old or more. All of them, however, involve an attempt to improve on nature, which usually turns out to be expensive and not always wise.

No Problem

Because of the expense involved in executing some of the more grandiose plans, some specialists have argued that nothing be done. In other words, they see nothing wrong with allowing the Caspian Sea to shrivel up and the Aral Sea to shrink into a salt marsh. Since almost all the other plans proposed involve expenditures of hundreds of millions of rubles, those who are prepared to see the seas contract or disappear question whether such sums are warranted. In other words, what harm will there be if water for the seas is cut off and used instead for irrigation or industry? The geographers V. L. Shul'ts and S. Iu. Geller make just such calculations (Shul'ts 1968, pp. 490–491; Geller 1962, p. 90). In their view the harmful effects of letting the Aral disappear would be—
1. A slight impact on the climate;
2. Fishing will decrease;
3. Muskrats, which provide valuable furs, will disappear;
4. Reed growth in the river deltas will disappear;
5. The Aral Sea will no longer be used as a waterway;
6. There may be dust and salt storms.

By their calculations, the loss entailed would not be large. Shul'ts and Geller have calculated that the loss suffered by the disappearance of fish from the Aral Sea will amount to 40 to 60 million rubles ($44 to 67 million) a year (Geller 1962, p. 90). The inability to use the Aral Sea for transportation will cost about 10 million rubles ($11 million).

Somewhat more extensive calculations have been prepared for the Caspian. Based on the estimate prepared in 1959 by the Oceanographical Commission of the Academy of Sciences of the USSR, the cost of relocating ports and channels in the Caspian would cost at least 100 million rubles ($111 million) a year (Apollov et al. 1963, p. 11). At the same time the revenue of the fishing industry was also estimated to have fallen by 100 million rubles a year (Apollov et al. 1963, p. 336; Geller 1962, p. 61). No one knows precisely how many other costs have been incurred. The geographer, S. N. Bobrov, estimates, however, that every time the Caspian drops one meter, it costs 400 million rubles ($444 million). As opposed to the loss of the seas, the gains that would follow from diverting the water flowing into it are as follows:

1. Instead of the useless evaporation of two-thirds of the flow of the Amu Darya and Syr Darya which would otherwise take place in the Aral Sea, eight million hectares of cotton land could be irrigated (Kes' 1967, p. 88). If in addition Lake Balkhash is drained and its Ili River harnessed for irrigation, another 5–6 million hectares of land could be put under cultivation. (Kes' 1967, p. 88;

Shul'ts 1968, p. 491). Some specialists calculate that since the yield from one hectare (2.47 acres) of irrigated land is about 1,000 rubles ($1,110) a year, this could add revenues of another 8 to 14 billion rubles ($9 to $15.5 billion) a year, of which about two-fifths would be profit (Geller 1962, p. 90; *Kom Prav* 11/22/68, p. 2).

2. The dried-out bed of the Aral, like the Gulf of Kara-Bogaz-Gol could serve as a valuable source of salt and other minerals.

Adding this all up, Shul'ts and Geller conclude that it would be wiser to let the Aral and Caspian dry up or shrink and use the water elsewhere. As they see it, diverting the water for irrigation would benefit the total economy by several billion rubles, whereas the loss of the Aral and the shrinking of the Caspian would cost less than one billion rubles. Accordingly, they see no immediate or convincing need to replenish these seas.

Localized Solutions

Most specialists, however, feel that the Caspian and to a lesser extent the Aral are worth preserving. Reconstituting a sea is obviously not simple or cheap. Where does one find more water when the existing stock of water is already in short supply and the surrounding area is very dry? The logical solution is to find an area with a surplus of water and bring some of this surplus in. As we shall see, this is in fact what many have proposed, even though "bringing it in" may mean transporting it several thousand kilometers. But recognizing the cost and ecological consequences of such a massive operation, others have suggested approaches that are more localized in character and which focus on redirecting the water-use patterns in the immediate area.

The most readily agreed upon solution is to concentrate on reducing pollution. That, after all, was a basic purpose of the October 1968 resolution on the Caspian Sea. Implementation of pollution controls would probably do little to halt the fall in the level of the sea, but it would prevent the further destruction of sea life to the extent that it stems from deteriorating water quality. To do this, water treatment installations would have to be built along the Ural and especially the Volga Rivers. Similar steps would have to be taken by the cities and installations along the Caspian. Improved treatment control by oil drillers, refineries, and shippers would be particularly important. But as everywhere else in the world, the Soviets have found that improved treatment does not come cheaply or without resistance. No matter which solution is finally adopted, pollution control seems to be an essential prerequisite. In any case, improved water treatment would generate less ecological disruption than the other plans.

In a similar vein, suggestions have been made to offset the fall in the flow of nutrient to the sea. As mentioned previously, this is caused by the erection of dams that block the movement of silt. One report shows that the phosphorus content of the Caspian, which used to be 100 milligrams per cubic meter of water, has fallen to 13 milligrams. The plankton, therefore, have less to grow on which in turn hurts fish and plant life. To make up for this deficiency, one specialist suggested pouring 3,000 tons of superphosphate into the Caspian (*Sov Ros* 7/14/70, p. 4). Ironically, the production of this fertilizer would use up some of the electrical power created by the dams that blocked off the silt-flow in the first place.

Since the loss of the Caspian fish is a major cause for the concern about preserving the Caspian, another localized solution is to increase the volume of artificial breeding. Many fish farms and artificial ponds have already been built and more have been called for. It is hoped that this will compensate for the loss of the natural breeding areas (*Sot In* 7/26/70, p. 2). But it is likely that the artificial breeding of fish and the feeding of the sea will never suffice to compensate for what used to happen naturally. Probably Russia's production of caviar will continue to diminish. One solution to this has been to develop synthetic caviar. In fact some initial success has been reported with a milk protein (*Sot In* 11/15/70, p. 3). As long as nothing happens to Soviet cows, this may suffice.

Some planners have proposed "a more imaginative scheme," the construction of a dam through the entire middle of the Caspian (Bobrov 1961, p. 56). According to the geographer, S. N. Bobrov, this would make it possible to maintain the shallow northern part of the sea even though less water flows into the Caspian. The southern part of the sea, which is deeper and suffers less from a fall in water inflow, will then be allowed to drop by one to two meters. A fish gate of some sort would be included to provide for the movement of fish across the

Caspian. Assuming the money would be available for such a massive project, there would certainly be a question of how Iran would react to such an undertaking. Iran also relies on the Caspian for a good portion of its fish and caviar. To the extent that the dam hampered the flow of fish from the breeding grounds of the north and to the extent that the southern part of the sea was made to bear the full burden of the diminished inflow of water, the Iranians presumably would oppose such efforts.

Since a dam across the Caspian is sure to create international if not ecological complications, the scientist I. I. Stass has argued instead for bringing in water from the Black Sea (Stass 1970, p. 24). At one time the two seas were linked together, so presumably this should not involve too much of an ecological disruption. It would also spare the fresh water of the Volga for irrigation. Stass suggested that the linkup could be carried out by using the Kum-Manych trough, which once served as a strait connecting the Sea of Azov to the Caspian.

Critics of this plan fear that the inflow of Black Sea water would throw off the ecological balance of the Caspian. Because the Black Sea is saltier than the Caspian, the inflow of this water might adversely affect fish life in the Caspian, which, after all, is the main purpose of the whole project (Bobrov 1961, p. 54). Moreover through the years, the Caspian has fallen 28 meters (92 feet) below the Azov so that the engineering requirements of this venture would be complicated. Even if everything could be worked out, some conservationists fear that this   /

plan may save the Caspian at the expense of the Azov. The high salt content of Black Sea water would be hard enough on the Caspian, but it would be devastating for the Azov which has also been severely damaged in recent years. Because of a reduced inflow of water into the Azov, which causes a higher salt content, the fish catch is only 5 to 10 percent of what it was a few decades ago (*Sot In* 8/15/70, p. 2; *LG* 5/17/66, p. 2; *LG* 5/17/66, p. 10). Some have even argued that 5 cubic kilometers of Volga River water be diverted from the Caspian to the Azov and that a dam be built between the Azov and Black Seas to prevent the further inflow of Black Sea water with its higher salt content into the Azov (*Trud* 2/11/70, p. 4; *SN* 2/16/69, p. 140).

As far-reaching as some of these proposals may sound, they are modest in comparison to the projects that have been given the greatest support and that to some degree have already been partially implemented. In their search for more fresh water for the area, planners in some of the scientific institutes have decided to go where water is actually in surplus. Under the circumstances, they were naturally attracted to the north and the Arctic Sea. As noted earlier, the purposeless discharge of all that potentially productive water into the inaccessible wastes of the Arctic tantalizes and frustrates government officials in the water-short areas of the south. Presently 80 to 85 percent of the river flows of the USSR move north into the Arctic or Pacific Oceans where seemingly they serve only 20 percent of the population (*Prav* 12/11/70, p. 3). Under the circumstances, what could make more sense than to use some of the water surplus of the north in order to eliminate the water shortages of the south? One part of this plan involves rerouting the Pechora and Vychegda Rivers in the northwest of the country, while a similar proposal involves the Ob and Yenesei in Siberia.

The Rivers of Siberia

The thought that it might be possible to "improve" on nature and reverse the flow of these mighty rivers did not originate entirely with Soviet planners. Prerevolutionary writers such as Ya. Demchenko advanced the idea of rerouting the Ob and Yenesei as early as 1880 (Shul'ts 1968, p. 491). Soviet planners have updated such proposals to take account of more advanced technology and geology, but the basic approach is the same (*Kaz Prav* 2/6/69, p. 2; *CDSP* 7/27/71, p. 12).

The project was to begin with the construction of a dam and storage reservoir at Tobolsk. Here the Irtysh and Tobol Rivers converge just before they go on to flow into the Ob (see Map 8.1). (*SN* 4/7/70, p. 6; *Business Week* 6/13/70, p. 43; *Kras Zvez* 1/17/71, p. 4; *Prav Vos* 1/9/68, p. 2; *SGRT* 1969, No. 3, p. 146.) In time the feeder area may be extended eastward by constructing a 1,500-kilometer (900-mile) canal. This would make it possible to add the waters of the Angara (which drain Lake Baikal) via the Yenesei to the reservoir at Tobolsk. Ultimately about 20 percent of the waters of the Ob and Yenesei could be diverted to this reservoir (*Kaz Prav* 2/6/69, p. 2). A chain of 12 pumping stations would force the water from the reservoir southward toward the

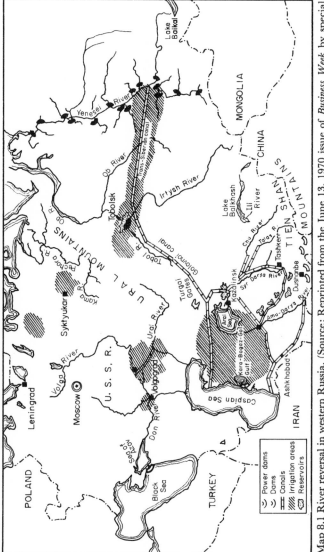

Map 8.1 River reversal in western Russia. (Source: Reprinted from the June 13, 1970 issue of *Business Week* by special permission. Copyright © by McGraw-Hill, Inc.)

north-flowing Tobol River, making it flow backward.
The water carried by the Tobol River would then move
through the Golovnoi Canal ultimately to the city of
Kazalinsk, located slightly east of the Aral Sea and near
the Syr Darya River. From Tobolsk to Kazalinsk is
1,500 kilometers (900 miles). More important than the
distance, this link necessitates a breakthrough into the
800-kilometer long (560 miles) Turgai Gates, the water
divide between West Siberia and the Aral-Caspian
Basin. If the canal is built, it will have involved a route of
over 3,000 kilometers (1,800 miles) via canals and river
beds through which 45 to 70 cubic kilometers of water a
year is to be moved (*Prav Vos* 1/9/68, p. 2; *SN* 2/17/70,
p. 79). The engineering work is equivalent to raising the
Missouri River and forcing it to flow backwards *over* the
continental divide so that it empties into the Pacific instead of the Gulf of Mexico.

Once the Turgai Gates have been breached, the rest of
the trip is almost all downhill. The water would move
into the Turgai River and on to the Upper Minbulak
Depression. There part of the water would be diverted to
a 4,500 square kilometer reservoir which would be
formed northeast of Kazalinsk and part of it would be
moved another 1,000 kilometers (600 miles) to the Caspian. The Upper Minbulak Reservoir would then feed
into a vast irrigation system which would be designed to
encompass almost all of Central Asia (Map 7.2). This
irrigation system would include the 110 kilometer (66
mile) Kazalinsk Canal which would feed into the Syr
Darya and the 870 kilometer (500 mile) Turkestan

Canal which would provide water for the irrigation of
the lower Syr Darya. The Amu Darya and Turkmenia
Canals would also be linked to carry water to the area
between the Amu Darya and Syr Darya as well as parts
of Turkmenistan. Several other smaller canals fill in the
system and theoretically should make it possible to irrigate
almost 25 million hectares (75 million acres) more than is
being done at present and without any further drain on
the Aral Sea (*SN* 2/17/70, p. 79; *Trud* 2/11/70, p. 4).

An alternative to the Tobolsk, Golovnoi, Kazalinsk
route is the diversion of the water from the Irtysh through
the Karaganda Canal. Under either approach, however,
the Aral Sea is unlikely to rise to its former level since
most of the Siberian water would be used for irrigation,
which promises a much higher economic return than
water used to fill the seas (*SN* 4/7/70, p. 7).

The Caspian Sea would be brought into this project
through the 650 kilometer (400 mile) Ust-Urt Canal (see
Map 7.2). The Ust-Urt would run between the Golovnoi
Canal and the Mangyshlak Peninsula on the east bank
of the Caspian (*Prav Vos* 1/9/68, p. 2; *Trud* 2/11/70, p. 4).
Once the water from Siberia arrives, the Caspian Sea
would not shrink as rapidly. With some of the urgency
removed, 5 cubic kilometers of the nonsalt water of the
Volga River could be diverted through the Don River
to the Sea of Azov in order to save it from the encroach-
ing salt water of the Black Sea.

The Rivers of the Northwest

While the diversion of some of the Siberian water to the

Caspian Sea may help, to most Soviet planners the fate of the Caspian is also dependent on another grandiose scheme—the rerouting of water from the northwestern part of the country. Although it now seems to have a somewhat lower priority than the Siberian River diversion, the benefits which it is claimed will arise out of it are virtually as far-reaching.

Like the Ob, Yenesei, Irtysh, and Tobol Rivers, the Pechora and Vychegda Rivers also carry their waters north to the polar regions where they serve "no useful purpose." As in Siberia, the inability to use such vast resources to many engineers and geographers only underlines the underutilization of Russian natural resources. It is not surprising, therefore, that the northwest, just like the Siberian project, has a prerevolutionary history. (See Map 8.2 for a version of 1784 which would link the Baltic and the Caspian.) Admiral Ribas suggested rerouting the Pechora and Vychegda to the Kama and then the Volga Rivers as early as 1789 (Shishkin 1962, p. 46). The current interest in reversing the rivers dates from 1933–1934 (Apollov et al. 1963, pp. 36–37). Survey and design work have continued periodically since that time. Several versions have been offered, but the most widely accepted plan was completed in March 1961 (*CDSP* 7/27/71, p. 12; *Kras Zvez* 11/4/70, p. 4; Apollov et al. 1963, p. 37). It was reworked and presented to Gosstroi (the Ministry of State Construction) of the USSR in March 1967 (*Trud* 3/19/67, p. 2). This variation was actually surveyed and plotted by demobilized servicemen during the 1960s but apparently the signal to begin

Map 8.2 Plan for a canal linking the Baltic and Caspian seas, 1784.

a massive construction program has not been forth-coming (*Kras Zvez* 11/4/70, p. 4; *NYT* 3/28/71, p. 1; *CDSP* 7/27/71, p. 12).

The general scheme that is now under consideration calls for the construction of a series of dams. Behind these dams, reservoirs would be formed which would raise the level of the rivers so that as in Siberia, the rivers would flow backward. Thus the first of a series of dams were to be built at Yaksha on the Pechora (Map 8.3). The Pechora River would then move south via the Pechora-Kolva Canal to the Kama and then the Volga (*Kras Zvez* 11/4/70, p. 4).[1] Further north, the Pechora would again be diverted to the Kama, this time by way of the Vychegda River (see Map 8.1). This would require a dam at Pokcha and Nivel Izhem, which would cut off the Pechora from its northern course (Apollov et al. 1963, p. 40). A canal would then carry the backed-up Pechora to the Vychegda River at Ust Kulom (*Kras Zves* 11/4/70, p. 4). Another dam at Ust Kulom would raise the combined waters of the Pechora and Vychegda and direct them into the Vychegda-Kama Canal, which would carry almost 40 cubic kilometers of water each year to the Kama (Bestuzhev-Lada 1969, p. 91; *Kras Zvez* 11/4/70, p. 4).[2]

In addition to the construction of the dams, there were also plans for a series of supplementary complexes that

[1] For a longer description of the alternatives, see Micklin 1969, p. 199.
[2] Not everyone agrees on just how much water will be diverted. Bazenkov says it will be 37 cubic kilometers whereas the report in *Krasnia Zvezda* says it will only be 8 cubic kilometers (Bazenkov 1968, p. 18; *Kras Zvez* 11/4/70, p. 4).

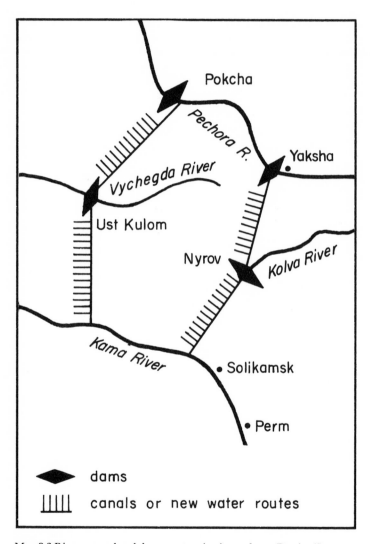

Map 8.3 River reversal and dam construction in northwest Russia. (Source: *Kras Zvez* 11/4/70, p. 4)

would provide water for hydroelectric power and irrigation (Micklin 1969, p. 199). Future possibilities provided
for the diversion of additional water from the Kubinskoye, Vozha, and Laga Lakes as well as the Sukhona
and Onega Rivers (*SN* 4/7/70, p. 8). Another variant
would reverse the flow of Lake Ladoga and divert its
water from the Baltic Sea in the north to the Caspian.
The backers of this project argued that it might obviate
the need for rerouting the Pechora and Vychegda (see
Map 7.1) (Kuznetsov and Tikhomirov 1965, p. 88).

As with the Siberian project, the construction costs of
these various alternatives would reach into the billions
of dollars and would involve construction of 800 kilometers (500 miles) of new waterways and the movement
of 680 million cubic meters of dirt (*EG* 2/21/61, p. 3;
Micklin 1969, p. 213; Sarukhanov 1961, p. 57; Apollov
et al. 1963, p. 42). It was claimed that such expenses
were warranted, however, because of an anticipated payback period of four years (*Kras Zvez* 11/4/70, p. 4). After
several postponements and revisions, including one that
eliminates the link up with the Vychegda River so that
only the Pechora, Kolva, and Kama are joined together,
expectations were in 1970 that the project would be
completed by the late 1970s (*Kras Zvez* 11/4/70, p. 4;
SN 4/7/70, p. 8; Detwyler 1971, p. 303; *CDSP* 7/27/71,
p. 12).

With the publication of the Ninth Five-Year Plan for
1971–1975, it seemed unlikely, however, that the northwest and Siberian projects would be completed as soon as
some planners would like. The Five-Year Plan contained

no evidence that full funding had been authorized as
at one time it was thought it would (*Sot In* 2/14/71,
p. 1; *NYT* 3/28/71, p. 11). In fact, an initial version of
the Ninth Five-Year Plan provided for the construction
of a dam on the lower Kama River (*Prav* 2/14/71, p. 4).
Authorization for this dam was omitted from a subsequent version of the plan (*Prav* 4/11/71, p. 4). Nevertheless as is frequently the case, the planners and engineers
on the projects have long time horizons and are normally
content to move piece by piece.

Major portions of both projects have already been completed. That at least is how the numerous dams that
have been built along the Volga and Kama have been
treated (Bestuzhev-Lada 1969, p. 89). One report indicates that the Volga basin will be linked up to water
from Northwest Russia by 1985 (*Sot In* 8/13/71, p. 4).
Similar work is progressing gradually in Western Siberia,
as we saw, along the 500-kilometer Irtysh-Karaganda
Canal (*Kaz Prav* 2/19/71, p. 2; Armand 1966, p. 79).
Instead of flowing north, water from the Irtysh now goes
to Kazakhstan (Bazenkov 1968, p. 17). This canal
breaches the Irtysh River near its origins at Yermak and
will soon provide additional water for the area around
Karaganda (Shabad 1969, p. 294; *Sot In* 4/29/71, p. 4).
Ultimately the river is again to be tapped at Tobolsk just
before it flows into the Ob. The Irtysh-Karaganda Canal
at Yermak goes a distance of almost 500 kilometers (300
miles) and is designed to carry 2.5 cubic kilometers of
water a year. Twenty pumping stations are to be built
so that the water can be raised a height of almost 1.5

kilometers (1 mile) (Bestuzhev-Lada 1969, p. 89). Work has also begun on a system of feeder canals. The Amu-Bukhara Canal is being expanded and more of the Amu-Darya water is being diverted from the Aral Sea on the assumption that ultimately the water will be supplemented with supplies brought in from Siberia (*SGRT* 1970, No. 6, p. 11).

Despite the completion of a considerable portion of the design and survey work, actual construction has moved erratically. In at least one case, this turned out to be most fortunate. Vast quantities of oil were discovered near Ukhta, one of the areas that was slated for flooding as a reservoir (Shabad 1969, pp. 20–22).

The Wayward Engineers

The flooding of some of the Soviet Union's most valuable oil reserves is a concrete instance of how improving on nature may generate costs that are greater than the benefits. Yet as some of the more imaginative planners have noted, even if the area had been flooded, the oil ministry could always drill over the water. As it turns out, this may have been one of the least complex effects of the proposed rerouting of the rivers.

Though some oil land has been spared, vast quantities of other expansive property are likely to be destroyed. This includes not only farm land, but towns and promising deposits of raw materials. According to most Soviet critics of flooding, destruction of vast areas of land in the USSR has been a problem in the past and is likely to continue as one as long as land and raw materials are

undervalued. Until the July 1967 price reforms, land
and underground raw materials were treated virtually
as free goods. A price is now charged but usually it does
not adequately reflect the economic values involved
(*Trud* 6/28/70, p. 2; Sukhotin 1968, p. 29). Similarly,
since 1966 dam builders have been obligated to see that
new lands are put into use to replace any land flooded
in the process of construction. But without an appropriate
charge for land such laws are not likely to generate much
response. ("Otsenka prirodnykh resursov" 1968, p. 51;
NYT 4/9/71, p. 2). Thus reservoirs or dam storage areas
in the USSR in 1970 covered 12 million hectares of
which 6 million hectares consisted of land that had been
or could have been used for agricultural crops (*Sel Zh*
7/14/70, p. 3; *CDSP* 12/22/70, p. 12). Valuable timber
and sometimes oil land has been flooded and sacrificed
in the same way (Gordeev 1963, p. 51; Armand 1966,
p. 81). As of 1965, 5.7 million hectares had been flooded
because of dam construction and other activities of man
(Armand 1966, p. 81).

Some proponents of the river reversal program argue
that even though flooding does occur, diversion of the
Siberian rivers may still result in a net increase in arable
land. Since many of the river basins in Siberia are already
severely waterlogged or marshy, these specialists feel that
removal of this surplus water to the south will help drain
40–60 million hectares of Siberian land (*SN* 2/17/70,
p. 79; *SN* 4/7/70, p. 7). To other Soviet critics such
estimates seem unduly high. These skeptics insist that the
building of these dams and reservoirs, particularly along

the lower Ob, "would have a disastrous effect on the general water regimen of the West Siberian plain" by causing an increase in the soil that is waterlogged (*SN* 4/11/69, p. 103).

Another by-product of the dam building and irrigation process is salination of the soil (Dolgushin 1969, p. 302; *Kom Prav* 9/16/65, p. 2). This is a particularly severe problem in dry Central Asia, especially where there is no suitable drainage of irrigated fields. According to Academician Gerasimov, salination "has now spread to a substantial part of our irrigated land." (*SN* 3/11/69, p. 103.) He goes on to point out that not only does salination lower the productivity of the land, it sometimes makes irrigated land completely useless. According to his data, the total area of land abandoned due to salination in Central Asia exceeds the area of newly irrigated land (*SGRT* 1968, No. 6, p. 448; Armand 1966, p. 40).

Concerned by what he feels to be excessive flooding, waterlogging, and salination, the respected Soviet ecologist, D. Armand, argues that it might be wise in some instances to sacrifice a bigger dam and greater electrical generating capacity for less flooded land and several low dams with smaller individual electrical capacity (Armand 1966, pp. 81–83). He, too, blames the failure to adopt a less destructive alternative on the absence of an adequate charge for land. With virtually no charge for land, planners have little need to make opportunity cost calculations to see if alternative projects which destroy less land would be cheaper. As evidence Armand points out that in contrast to practice in noncommunist countries, Russian planners and engineers seldom build their dams in mountain valleys where there is less good land to be engulfed (Armand 1966, p. 83). The reason for this is that dam construction in the mountains is usually more complicated and dangerous and therefore more expensive. Consequently as long as any land lost through flooding is considered a free good and not an element to be included in any cost estimate of alternative dam sites, it is natural that the engineers will select those sites that are the most accessible and where construction costs will be lower, even though this usually means the flooding of many more acres of land.

More realistic pricing of water would also reduce the flooding and soil salination that takes place. If water were more expensive, less of it would be squandered or wasted. Peasants would also use it more sparingly on their fields and more effort would be put into ensuring that irrigation canals were well constructed to prevent leaks. ("Otsenka prirodnykh resursov" 1968, p. 81; *Kom Prav* 9/16/65, p. 2.) Because water is so cheap it does not pay to spend much on the construction or the maintenance of the irrigation canals. As a result there is extensive evaporation and seepage. According to one estimate, 7 million cubic meters of water a year are lost through leaks and evaporation just from the Karakum Canal (*Kom Prav* 9/16/65, p. 2). L'vovich feels that lining the canals would increase the efficiency of the irrigation system by 50 to 100 percent. This alone would make it possible to irrigate 30 to 50 percent more land with the water supplies presently available (L'vovich 1962, p. 111). Careless irrigation practices are also a significant cause of salination. The combined effect of a price on water and a higher price on land might make it economically advantageous to cover and line irrigation canals so as to prevent evaporation and salination. If it is still decided to bring in water from Siberia, a proper price on water might make it worthwhile to bring it in through a system of pumps and pipes rather than through open ditches, canals, and dams with flooded reservoirs (*Kom Prav* 8/11/68, p. 2; *LG* 6/17/70, p. 11).

Another way to reduce evaporation would be to follow the advice of those who argue that the flow of water into the Gulf of Kara-Bogaz-Gol should be curbed. By blocking off the passage way into the gulf, some geographers claim that there would be a saving of 10 cubic kilometers of water a year in reduced evaporation (Vendrov et al. 1964, p. 32). This suggestion is opposed by industrial ministries which fear the closing off of the inflow of water and the resulting cessation of operations at Kara-Bogaz-Gol will make it more difficult and costly for them to obtain needed raw materials.

Juggling the natural environment around by reversing rivers and building dams often sets off other unanticipated reactions. It is not always easy to pinpoint, but dam construction down to bedrock and stream diversion seem to disrupt underground water flow. This could create chaos for regions hundreds of miles away which suddenly find their underground wells have dried up. (*SN* 3/11/69, p. 102; *LG* 6/17/70, p. 11; *Kom Prav* 8/11/68,

p. 2). Similar side effects are likely to follow the building of vast new reservoirs throughout Siberia and Central Asia. New and larger water surfaces are now exposed to evaporation. Several ecologists feel this could give rise to entirely new weather and rain patterns, not to mention what changes it might have on the flora and fauna of the area (*Kaz Prav* 2/6/69, p. 2; *SN* 3/11/69, p. 102; Gerasimov 1962, p. 5; *SGRT* 1964, No. 3, p. 60; Mezentsev 1964, p. 24). Among other possibilities, the creation of broad unobstructed spaces on the surface of the new lakes and reservoirs could easily lead to severe wind storms (*SGRT* 1965, No. 10, p. 3). This is already a problem in many parts of the USSR. Much effort of Soviet conservationists has been directed to the creation of barriers such as trees. Flooding and creation of new water bodies is a step in the opposite direction.

Equally uncertain is the effect of this geographical facelifting on the Arctic Ocean. Only a few ecologists have dared to question what might happen if the flow of these large rivers were to be cut off. Although their conclusions are sometimes contradictory, the one thing they agree on is that enormous forces may be set in motion which could affect the whole earth (Gerasimov 1970, p. 523).

According to one school of thought, the diversion of all that warm water from the Arctic would deprive the region of a major portion of what little moderating influence it now receives (*SN* 3/11/69, p. 105). Without water from these warmer regions of the country, the Arctic ice cap will spread. At worst this could mean a return to the ice age (*NYT* 7/18/70, p. 36; Gerasimov 1962, p. 37). An opposite but no less appalling theory has been advanced by Hubert Lamb, a British climatologist (*Sea Secrets*, p. 6). He fears that the diversion of the Ob, Yenesei, and Pechora would deprive the Arctic Ocean of about one-half of its freshwater supply. In his view, this water "keeps the top layer of the ocean comparatively fresh, so that it freezes more easily. If the supply was reduced or cut off, there would be large scale melting." Lamb contends that this could lead to a warming of the northern hemisphere. The Mediterranean climate would move northward and those areas that now have a Mediterranean climate would probably warm up so that they had a North African-type climate. Going on from here, one can speculate that such a warming effect would cause a melting of the ice cap and the subsequent flooding of vast regions of the earth.

Equally upsetting is the hypothesis of Dr. Raymond L. Nace of the United States Geological Survey. He asserts that such massive river reversal projects, be they in the Soviet Union or the United States, could affect the rotation of the earth (*NYT* 2/13/70, p. 61). By shifting such a vast weight from the North Pole to the Equator, the spin of the earth could be retarded. Furthermore, just as the rotation of a wheel is affected by moving a small weight around the rim, so the earth might develop a wobble. Baseball pitchers and umpires have long known that by roughing the seams of the ball or a large enough section of the surface, the flight of the ball can be affected significantly. Altering the earth in such a profound way, however, can hardly be considered a game.

River reversal and dam construction have ramifications not only upon nature but upon our social and economic activity. Initially the prospects and often the results seem to be positive. Automatically, increased utilization of the area is of course desirable, especially if the full social costs are not reflected in the costs of operation. More often than not, however, the time quickly arrives when the resources have become overtaxed again so that even greater expenditures are required to remedy the newest exigency. This is comparable to what we can call in the United States the "build another highway" syndrome. The more highways that are built, the more cars crowd onto those highways. In the case of water usage, the misdirection of resources is intensified because the newly available water is usually heavily subsidized either as an outright gift or as a by-product of electricity generation. Moreover once such projects become part of a hydroelectric complex, electricity generation usually take precedence over the other water needs. Thus, as we have seen, instead of being released at times most appropriate for fish breeding, water is conserved behind the Volga dams to ensure the uninterrupted flow in electrical generation (*Trud* 3/19/67, p. 2).

The misuse of water resources is not unique to the USSR. Exactly the same kind of action and reaction takes place in the southwestern part of the United States. Because of the shortage of water, states such as Arizona apply enormous political and other pressure to bring in more water from the north. Dams are built. Cheap water comes in, which in turn attracts more industrial and agricultural activity, so that water becomes scarce again. Thus just as in Arizona, the diversion of water in the Soviet Union from the Dnepr River to the water-short Donbass Region 550 kilometers away has proven to be only a short-run solution. New industry in the Donbass has more than absorbed all the extra water that was brought in by this canal, which has cost $600 million (*Trud* 11/12/66, p. 2; *VST* 2/71, p. 3). Of more immediate relevance here, the same criticism has been made about the Irtysh-Karaganda Canal (*Trud* 11/12/66, p. 2). In fact, the whole river reversal effort is likely to lead to even greater misallocation of resources. There are many economists and geographers who charge that as long as the rivers are increasingly diverted for irrigation and industrial use, the Aral and Caspian Seas will never return to their normal levels (Zhuvkovich 1967, p. 110). New and cheap water supplies attract new and thirsty users and, like highway construction, it is seemingly impossible to keep ahead of demand.

Despite such ominous warnings, there are still many engineers and economists who feel that these water diversion projects should move ahead, and indeed they support the construction that is already underway. Since the natural consequences of their activities are rather frightening, the river movers often try to find economic arguments to substantiate their cause. They calculate how much agricultural production can be increased through irrigation but not how much land will be removed from agricultural use. As we have seen, cost benefit analysis can be used to prove almost anything, including the benefits of allowing seas to dry up. Moreover no matter how many billions of rubles a particular project may cost, not only is the economic benefit usually shown to be higher, but the calculations are made to show that the project has a very short payback period (*EG* 2/21/61, p. 3). In fact it sometimes seems that the project can be made to pay for itself almost before the work is even completed.

  Calculations of this sort should be as suspect in the Soviet Union as they have become in the United States. In addition, in the Soviet Union, as in the United States, actual construction expenditures somehow almost always end up exceeding the original estimates. Furthermore, even where calculations are conducted with an extraordinarily conservative perspective, the cost benefit analysis is misleading because social or external costs are so hard to quantify. Some critics of these grandiose plans have come to recognize the handicap they are working against and that they cannot include the psychic cost of losing the water for recreation and other nontangible uses, but there is little they can do to counter such reasoning (*Trud* 11/12/66, p. 2). One of the few acceptable arguments the critics can muster is that while river reversal may benefit the Central Asian regions, it will do so only at the expense of Western Siberia (Vendrov 1964, p. 23; *LG* 6/17/70, p. 11). This accusation is one national officials if not planners must heed just as they listen when the San Franciscans complain about the water-napping proclivities of the Los Angeleanos. Regional pride and self-interest is one of the few forces that is an acceptable challenge to "progress." Unfortunately in the case of the Siberian rivers, even regional rivalry does not appear to be a potent-enough barrier.

Conclusion

Nature is not always a perfectly balanced mechanism. Periodically it brings forth its own disasters which have nothing to do with man. What concerns us here, however, are those instances where, because of man, disruptive or destructive tendencies are created or accelerated where otherwise there would have been no difficulty. As Engels implied, once the initial damage has been done, the iterative effects do not always dampen down. Diverting the Volga, Syr Darya, and Amu Darya has given rise to problems that seemingly can only be solved by generating new difficulties.

Critics of such efforts should not be regarded as Luddites who are opposed to all technological change or experimentation. They are not necessarily against "progress," nor do they seek a return to the "good ole days." What they are concerned about is that technological prowess is so often embraced without skepticism. They are especially wary of engineers who insist that no matter how serious the trouble, they can provide a solution. In a world that dashes from one technological revolution and solution to another, some conservationists now feel that placing blind trust in science and technology is too dangerous a policy to follow.

Some of these dangers could be reduced if somehow social costs could be more accurately anticipated and reflected. The hazards are further aggravated in the USSR where all costs, not just social costs, are misrepresented. Together this intensifies what we can call "Tsuru's complaint." Shigeto Tsuru of Hitsubashi University in Tokyo is perplexed by the tendency of man to spend large sums of money in order to create new facilities when often existing resources are available and could be used just as well. It seems especially irrational when environmental disruption is generated in the process (Tsuru 1971, pp. 10, 14). Thus, he wonders why new land must be reclaimed when existing land is underutilized or unutilized a few miles away. In the Soviet context, Tsuru would presumably ask why new lands must be irrigated when present lands are being used inefficiently or unproductively, especially when good land is flooded in the process of building the reservoir. If the existing resources were properly used, productivity would increase, so that presumably such a massive redesign of nature might be unnecessary.

The propensity to authorize enormous public works schemes is likely to increase in the future for at least two reasons. First, each day new and more advanced machinery is invented which increases the scope of man's capabilities. His potential for change steadily becomes greater. Second, the desire of the developing countries to catch up with the ever-moving developed countries becomes greater as the distance between the developed and developing increases. When massive public works projects are presented as one-jump solutions for closing the gap, the temptation becomes irresistible.

Decisions of this sort often have ramifications beyond the borders of one individual country. The example of what could happen to the ice cap of the Arctic or to the change in the weather pattern of Asia are some examples. The fall of the fish catch in the Caspian, which also affects Iran, is another instance. Some planners simply do not care if other people suffer while they fulfill their self-assigned destiny. Two scientists, I. A. Kuznetsov and F. K. Tikhomirov suggest that it would be far wiser to reroute the water leaving Lake Ladoga from its present destination *outside* of the country so that it would flow instead south to the Caspian *within* the USSR (Kuznetsov and Tikhomirov 1965, p. 88). According to Kuznetsov and Tikhomirov, it seems unfair that the water, which, after all originates in the USSR, should be allowed to flow outside the country where it does the USSR no good. The potential impact of such a change on the White Sea and Scandinavia does not concern them.

National selfishness, however, does not mean that international cooperation will necessarily obviate all the problems. On the contrary, if several large countries in the world started to cooperate on such projects, the hazards could be just that much greater. For example the Japanese are already worried that the United States and the Soviet Union will decide between themselves to build a dam between Alaska and Siberia across the Bering Strait. Colder Pacific Ocean water would then be blocked, which should then induce a larger inflow into the Arctic of warmer water from the Atlantic Ocean. To the Japanese this threatens to throw off their whole ecological and environmental balance. Nonetheless, not only does the Russian scientist Bestuzhev-Lada urge the construction of such a dam across the Bering Strait, he also dreams excitedly of building another 400-kilometer (250-mile) dam along the northern shore of the Black Sea from the Danube to the Perekop River (Bestuzhev-Lada 1969, p. 93). The water that normally flows into the Black Sea could then be used for irrigation, while water from the Mediterranean Sea would flow into the Black Sea. In this way, the Black Sea would also become warmer. The warming effect of the dams across the Bering Strait and the Black Sea would be supplemented by another dam Bestuzhev-Lada proposes for the Baltic Sea. Together these projects would cause the ice cap to melt. Bestuzhev-Lada feels these projects would provide countless benefits for mankind, all at a cost of only 11 billion rubles ($12 billion). (Bestuzhev-Lada 1969, p. 95.) Unfortunately, Bestuzhev-Lada sees one major obstacle to the implementation of such a grandiose scheme.

"To carry out these dreams at the present time, there are no other obstacles except the politics of the imperialist powers which keep up international tensions."

This is probably the first convincing argument for maintaining imperialism and continuing the cold war!

9 Advantages of the Soviet System and International Implications of Soviet Policies

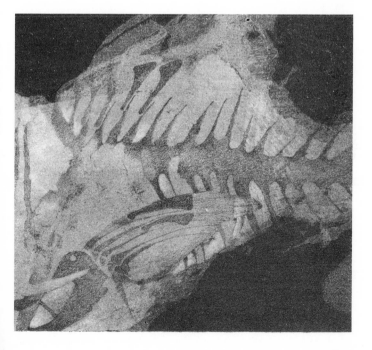

Having analyzed the forces that brought about environmental disruption in the USSR, it seems appropriate that attention now focus on the advantages the Soviet system may have for coping with environmental disruption. The initial stimulus for this study stemmed from the expectation that the Soviet system would offer guidelines and experiences which could be of help in other countries. While the Russians do in fact have something to teach us, regrettably the lesson is a short one and not that helpful. Would that it be that someday the Russians will be able to show us the way toward not only a better theory for the control of environmental quality but toward an actual implementation of that policy. Until that day arrives, however, it becomes more and more important to pay heed to what the Russians are actually doing, because the effects of environmental disruption increasingly extend beyond the borders of even large countries. Consequently John Donne's statement, "No man is an island, entire of itself," might be read "No island is a planet, entire of itself." Today what happens on islands, not to mention large countries such as the USSR, often spills over into the world's ecological system and may affect the entire earth. We conclude then with a discussion of how and where Soviet activities affect the rest of the world and what can be done to encourage international cooperation in reducing environmental disruption.

Advantages of the Soviet System
One of the major arguments throughout this study has

been that the concentration of economic and political power in the hands of the Soviet state can be and has been a major factor in the creation of environmental disruption. In the Soviet government's drive toward industrialization and economic growth all too often there has been no person or group around with any power to stand up for the protection of the environment. Until the point is reached when environmental disruption causes other state interests, especially manufacturing and agriculture, to lose as much as those in favor of greater exploitation of the environment stand to gain, environmental quality in the USSR is in a very fragile condition. Still, this argument can be turned around. For whatever reason, assume that the state suddenly decrees that all activity in a certain area must cease. With its enormous power, the Soviet state has the potential for being a most effective tool for conservationists.

Preserves An example of where the state has used its power in this way is in the creation of natural preserves. Although the first such preserves were created before the revolution, we saw how the creation of additional reserves was encouraged during the very earliest days of Lenin's rule (Bannikov 1968, pp. 89–96). Just as the lack of private land ownership can facilitate state expansion and devastation, so the absence of private land owners can make it easier for the state to act when it decides to set aside lands as natural preserves. There is no need to spend time or money to defeat lengthy court litigation brought by private landholders as happens when the U.S. government tries to establish a national park on

Cape Cod or other national parks (*WSJ* 4/29/71, p. 34). It may take the USSR some time to decide to act, as we saw at Lake Baikal where no action as yet has been taken on the expansion of the Barguzinskii Preserve, but when it does move, the state can encompass large tracts of land in a single decree. As of 1970 the Soviet government had set aside nearly 8 million hectares (20 million acres) for such preserves (*Sot In* 8/27/70, p. 4).

Nevertheless Soviet preserves have been criticized because they sometimes have not been properly protected (Armand 1966, p. 134; *Prav* 11/21/67, p. 3). In fact the number and size of the preserves have actually been reduced since 1950. While now there are a total of 90 such preserves, at one time there were as many as 128 (Armand 1966, pp. 180, 182). Furthermore, Professor Armand complains that the Soviet Union lags far behind all other developed countries in the world in terms of the percentage of the total land it has set aside for public parks. From a peak in 1950 when 0.56 percent of Soviet land was devoted to public parks and preserves, the percentage had fallen to 0.28 percent in 1966. This compares with 1.5 percent in the United States, 1.5 per cent in Canada, 4 percent in Great Britain, and 6 percent in Japan (Armand 1966, p. 182; Bannikov 1968, p. 96; U.S. Bureau of the Census 1970, pp. 192, 198, 200).

This decrease in the number and size of preserves has taken place despite the existence of a law that decrees that the territory in such preserves is to be removed from economic use forever (Armand 1966, p. 180). Professor Armand mentions several instances where Soviet industrialists have moved in as soon as the preserves were abolished. On occasion even though the preserve is not abolished, industry may still encroach on its sacrosanct areas. A paper plant at the mouth of the Volga near the preserve outside of Astrakhan was allowed to harvest reeds in the otherwise protected and swampy areas (Armand 1966, p. 119; *LG* 3/30/65, p. 2). There have been numerous other abuses reminiscent of similar misuse of American natural preserves. Soviet forest preserves have been used for timber, for highway construction, and even for the private hunting and social pleasures of the Minister of Agriculture, V. V. Matskevich, and his personal friends (*Kom Prav* 8/12/68, p. 2; *Kom Prav* 10/14/70, p. 2; *CDSP* 7/7/70, p. 5). It is difficult to see how such abuses can ever be totally eliminated given the nature of bureaucracies and bureaucrats and the fact that Minister Matskevich himself is now in charge of safeguarding the sanctity of several of the preserves (*CDSP* 7/7/70, p. 5).

Central Heat and Hot Water Soviet policies regarding the generation of central heat and electricity have been considerably more successful. Because the state owns all the utilities as well as most of the buildings, the Russians have stressed the installation of centrally supplied steam from the Teplovaia Elektrotsentral (TETs). As we have seen, heating and hot water are provided by central stations, which makes possible more efficient combustion and better smoke control than would be achieved if each building were to provide heat and hot water for itself. There were no such installations prior to the revolution, but as of 1971, 65 percent of all buildings in Moscow were centrally heated and supplied by the TETs. By 1980, it is expected that Moscow will have 90 to 92 percent of all its buildings centrally supplied with heat and hot water. Reflecting their lower priority, Leningrad and Kiev buildings obtain only 40 percent of their heat this way, although future plans also provide for more such service in these cities as well.

Ultimately it is planned that all electric power generating stations in metropolitan areas will also have steam generating capabilities. In the past, even though steam was a by-product of electric power generation, the electric power station would be designed only to provide electricity and a steam plant would be designed only to provide steam. As of 1970 there were already 12 joint stations in operation. Their capacities ranged from 600,000 kilowatts to 1 million kilowatts. A larger plant with a capacity of 1.5 million kilowatts is under construction and a 2-million kilowatt plant is to be designed by 1980 (Zinger 1970, p. 16).

Some American cities such as Boston and New York have similar facilities available. Such service, however, is optional in the United States. If a landlord so chooses, he can generate his own heat and even his own electricity. This defeats the purpose of restricting all chimney effluent to one centrally controllable source. Greater use of the TETs system of the USSR by U.S. utilities would also go a long way in reducing thermal pollution. At the same time, more power generating plants should also be transformed into steam generating plants as well. By taking the heat and steam thrown off in the process of electrical generation and using it for heat, air conditioning, and hot water, the power plants would no longer need to dump their hot water wastefully into nearby water courses. Moreover, expecially where there is a high density of office or hotel buildings as in New York City or Miami Beach, new fuel sources need not be expanded to produce the heat, cool air, and hot water as is presently being done.

Not everyone in the USSR, however, is happy with the TETs system. There are complaints about excess heat loss and operating malfunctions, so that when one part of the system breaks down, an entire section of the city is affected (*VST* 5/70, p. 38; *Prav* 9/24/69, p. 3). Moreover there is sometimes difficulty in providing for the varying heat needs of different buildings and rooms in many of the TETs operations (Livchak and Kop'ev 1969, p. 21). Another difficulty is that occasionally the gains derived from concentrating the combustion in one large furnace and boiler are offset by the need to send steam and hot water long distances. There is a trade-off point where the loss of heat through the connecting pipes necessitates so much extra combustion that the effluent is actually increased by not having individual boilers and furnaces. Finally for the people living in the immediate vicinity of the TETs, the cleaner air and greater efficiency per BTU (British Thermal Unit) generated is offset by the fact that more BTUs are being generated in one location. While the city as a whole may have cleaner air, this may be little compensation for the neighborhood around the TETs, which will probably be worse off than if the traditional decentralized heat and hot water system were used. Despite these drawbacks, there are both ecological and economic possibilities here for all metropolitan areas in the United States which deserve further study.

Zoning and Plant Location When it chooses to, the state in the USSR can also exercise enormous power in such matters as plant location and zoning control. We have pointed out numerous instances of where these rules are violated, but they can also be used effectively. On occasion some green belt areas have been established and protected and some production processes altered to conform to environmental standards. Although a common complaint is that this power is not used enough if at all in most parts of the USSR, some plants such as the Novogor'ky Oil Refinery and the Voskresensk Chemical Combine in Moscow were closed down at least temporarily, because they were polluting the water (*LG* 6/13/70, p. 11; *Trud* 12/25/70; p. 2; Armand 1966, p. 69; Vitt 1970, p. 78). Still on the whole, powers for the prevention of environmental disruption in the USSR have been honored more in theory than practice (*Prav* 6/21/65, p. 2; *Med Gaz* 11/11/69, p. 11; Kolbasov 1965, pp. 43–45; *CDSP* 7/7/65, p. 14; *CDSP* 12/13/69, p. 14). In comparison, since 1970, when the mercury scare in the United States prompted much greater concern and the powerful Refuse Act of 1899 was rediscovered, environmental agencies in the United States have acted more vigorously in forcing factories to close down or halt their emission of toxic effluents than Soviet authorities.

De-emphasis on Consumer Goods For good or bad, there has also been much less emphasis on consumption and product obsolescence in the USSR. This downgrading of consumption is primarily due to the dominance of the state in making economic decisions and deciding on the allocation of resources and investment funds. Soviet political leaders and planners were determined to build up Soviet economic and military power to ensure the military and political security of the state before they would permit more than a minimal allocation of resources for the consumer. Critics of Soviet policy have argued that even with two world wars and a counterrevolution or two, 55 years of power is time enough to industrialize a country and provide more for the consumer. The Japanese have come from an even less developed state and nonetheless have managed to provide for their consumers. Nevertheless, given the past de-emphasis on consumer goods at least until the Ninth Five-Year Plan of 1971–1975, the Russians have less product proliferation and fewer goods. Therefore, they have less waste to discard. Moreover, the tendency to underprice labor in relation to consumer goods results in much more recycling in the Soviet Union than in the United States. This difference in American and Soviet price ratios of labor to consumer goods is partly due to the imposition of a turnover tax, which until recently averaged about 50 percent of the final price. It is also due to the relatively low productivity of Soviet labor, particularly service labor, in comparison to the United States.

Given the low priority for consumer goods and the unusual price ratio, Soviet consumer goods tend to enjoy a much longer life in the USSR, not because they are of higher quality but because they are scarcer. This not only reduces the complexity of the solid-waste problem, but it leads to resuse and recycling which reduces somewhat the consumption of the raw materials themselves. Because labor is so cheap in relation to most products, it pays to hire someone to collect wastepaper, old clothes, and scrap metal. There is no such thing as a nonreturnable bottle. Even vodka bottles are returnable with a deposit.

In the USSR as in other developing countries, one is struck by the large number of what appear to be antique automobiles in the streets. Because of the general scarcity of automobiles, it is worthwhile for the private driver to incur a high level of maintenance costs because generally there is no replacement available at any legal price. Normally a car will be repaired and reused regardless of its age, or it will be cannibalized into spare parts. The day when Moscow will be plagued like New York with 70,000 abandoned automobiles on its streets seems a long way off, especially when total automobile production in the whole country was slightly less than 350,000 in 1970. Unfortunately, Soviet consumers are probably considerably less enthusiastic about the shortage of consumer goods and automobiles than the ecologists and conservationists. Still, from the environmentalist's point of view, the fact remains that there are no disposable bottles and diapers and very few abandoned automobiles to worry about in the USSR.

While the Soviet price system has facilitated the de-emphasis on consumption, it has also led as we saw to the undervaluation of raw materials. Although such a system has created difficulties for the environment in the past, some day it could conceivably be used to the environment's advantage. Just as the Russian prices in the past have not reflected land and capital costs, so some future Russian pricing officials might decide that all Russian products henceforth will include some specified markup intended to reflect social costs. It would still be hard to determine just what social costs are attributable to which products, but implementation of this decision in the USSR should be easier to carry out than it would in other societies.

As in all such discussions it is sometimes hard to ascertain what can properly be attributed to the communist or socialist nature of the Soviet government and what is merely due to the sociological heritage and material endowment that the Soviet government inherited. In many instances it is a combination of tradition, endowment, and power which affects what happens. Without the existence of large deposits of natural gas, for example, the Soviet government would not have been able to order the phasing out of coal-fired furnaces and boilers so rapidly. If the USSR had to rely on imports for all its natural gas, the process would have been considerably more expensive or it would have taken even longer for the government to decide to make the transformation. As it was, the USSR was slow in recognizing the advantages of using more gas and oil in its fuel balance, but its large supply of fuel reserves did ultimately allow it to make the change. It is likely that in a few years the state planners will be able to provide enough natural gas so that gas will become the main source of fuel in the major cities and the only source of electricity and steam during the four smokiest months of the year.

Sewage Recycling Sewage disposal is another area where the Russians have conducted some noteworthy experiments. As was pointed out, sewage farms are a prerevolutionary practice and sewage for irrigation in the Ukraine was used as early as 1894 (Leporskii and Nazarov 1969, p. 98). Soviet authorities have continued and expanded this form of recycling. If freed of dangerous industrial wastes and used on the proper crops and with the proper care, the pretreated effluent of the municipal sewage system contains rich nutrients which can be usefully applied to the soil. When used carefully this way it can obviate a good part of the need for artificial fertilizers. In a sense, the return of human waste to the soil rather than the stream is nature's way. Returning sewage to the fields is a common practice in Asia and even in parts of Europe and Israel. Now the Metropolitan Sanitary District authorities in Chicago, Illinois, are conducting similar experiments with Chicago's sewage on farms about 160 miles downstate. Similarly for many years gardeners and farmers throughout the United States have made wide use of the fertilizer Milorganite, which is nothing more than the dried-out and treated sludge of the Milwaukee sanitary treatment plant. But while only one or two American cities have embarked on recycling programs of this type, the Soviets have recently gone much further. In 1938 only 233 hectares of land were irrigated with sewage in the Moscow Oblast (Tolstoi and Bondarev 1969, p. 73). By 1958 the total for Moscow had increased to something under 5,000 hectares, while the total for the whole country was 11,000 hectares (A. I. L'vovich 1963, p. 41). The program grew rapidly in the years that followed so that in 1970, 70,000 to 80,000 hectares (175,000–200,000 acres) were irrigated in this way and it was planned to add another 60,000 hectares (150,000 acres) by 1971 (*VST* 10/70, p. 37). Over one hundred Soviet farmers around such cities as Moscow, Kiev, Zhdanov, Odessa, and Kharkhov agreed to irrigate their fields with municipal sewage (Leporskii and Nazarov 1969, p. 98). Around Leningrad, where soil conditions are not suitable for irrigation with sewage effluent, the dried solid waste from the water treatment plant is frequently taken by collective farmers free of charge for some use on the farm. The reuse of both solid and liquid wastes has on occasion caused difficulties, especially if applied directly to food crops, but on the whole if used with restraint it does seem to be a natural and an effective procedure (Grin and Koronkevich 1963, p. 123; Kolbasov 1965, p. 85).

International Implications

If the Russians are unable to use the theoretical potential of their system for the improvement of environmental quality and if, as seems the case, environmental disruption in the USSR should continue or become more serious, the ramifications increasingly will spread beyond the confines of the Soviet border. Less and less will environmental disruption in the USSR be a purely internal matter. Of course, the same holds true for environmental disruption in the United States or any other country. Environmental disruption, even in a country as vast and seemingly isolated as the Soviet Union, becomes everybody's business.

International involvement pertains especially to ecological systems that are shared by the USSR and some other country. One of the best examples is the Baltic Sea. It is hard to judge exactly which country bears how much responsibility for what has happened to the Baltic. Even the clean and careful Swedes have done their share toward its degradation (Fonselius 1970, p. 2). But whereas the Swedes and now some of the other Scandinavian countries have come to realize the error of their emissions and have taken bold corrective steps, the Russians have moved ever so slowly. All the seaports in the Baltic Republic of the USSR as well as Leningrad are important centers for the storage and refining of oil. Between the oil tanks and depots, tons of oil and ballast have escaped or been discharged into the Baltic over the years from every major port in the area (*EG* 9/67, No. 37, p. 38; *Sov Ros* 8/12/67, p. 2; *Sov Lit* 7/24/69, p. 2; *Sov*

Lat 3/22/68, p. 2; *Sov Lat* 12/14/69, p. 2; *Sov Est* 8/19/70, p. 2; *SN* 2/11/69, p. 63). Other harmful effluent comes from industries such as milk and meat processing, liquor distilling, paper and lumber mills, and chemical plants located along the shore (*Sov Lat* 3/22/68, p. 2).

With the discovery that the Baltic is near death, the Soviets have begun to take corrective action at home and to participate in joint efforts with their neighbors to save the sea. Toward this end the Russians have stepped up efforts to provide facilities for the discharge of ballast and the processing of sewage. For several years they held back from cooperating with the other Baltic states in cleanup efforts. Recently, however, they have shown a much greater willingness to participate and even to lead by calling for joint conferences.

The actual and potential deterioration of other jointly shared water bodies is something that concerns other neighbors of the USSR as well. Iran, as we saw, is seriously affected by what happens to the Caspian Sea and the flow of the Volga. Turkey, Rumania, and Bulgaria border on the Black Sea. Even where countries do not share a common water front with the USSR, they may be adversely affected. Because the Russians have carelessly overfished many of their own waters, they have moved out in a systematic and all-encompassing way to fishier fields elsewhere around the world (*Prav* 1/8/69, p. 3; *Prav* 3/25/69, p. 3; *Prav* 7/29/69, p. 2). As an indication of the effect of Soviet domestic practices, the total fish catch of fish within the Soviet Union rose by only 25 percent from 1913 to 1967 (*LG* 9/25/68, p. 10). Even this increase, however, was primarily due to a switch to a higher proportion of smaller and less prized fish in the total catch (*LG* 3/5/69, p. 10). In some areas such as Lake Ladoga, the water bodies have been virtually fished out. In other water bodies such as Lake Onega, and, as we saw, the Caspian and Azov Seas, the catch of valuable fish has fallen off drastically (Armand 1966, p. 135; *Prav* 1/8/69, p. 3). At the same time that the domestic catch suffered, the overall fish catch from 1913 to 1967 rose by 600 percent (*LG* 9/25/68, p. 10). This means that almost all the increase was due to fish caught in international waters and that the catch of international fish rose by a factor that was considerably larger. As indicated by the complaints of those who have experienced the impact of the Soviet fishing fleet, Russian fishermen generally show even less concern for the fishing grounds outside their own country than they have demonstrated for their own fishing grounds. There is fear, therefore, that the Russians may deplete some of the best and heretofore unexhausted fisheries of the world (*BG* 4/4/71, p. B-3; *SN* 12/9/69, p. 129).

We have seen that the scheme to reverse the north-flowing river is another instance where the whole world may be affected by actions taken within the boundaries of the USSR. As remote as they are, changes made in the Barents and Kara Seas and the Arctic Ocean affect the weather and well being of the whole northern hemisphere. In the same way, the laying of a pipeline across the perma-frost of Siberia toward Japan should arouse many of the same concerns as the contemplated American pipeline across the permafrost of Alaska. In both projects the oil pipeline could crack, which could spew oil over vast areas of sensitive and not easily regenerated plant and animal growth.

Like other nations, the Soviet Union has developed industrial processes in recent years, the impact of which will have a profound international effect. Even though there is a treaty banning the above-ground testing of atomic devices, the Soviet Union like the United States is rapidly moving ahead with atomic power generation (Medunin 1969, p. 31; *Trud* 12/19/68, p. 4). There is still much to be learned from such operations. Furthermore there is no reason to believe that the Russians have been any more successful than anyone else in the world in guaranteeing that there will be no low level leaks of radioactive elements into the air or water. Because of its long half-life, such radioactive material can easily work its way into international air and water basins.

Another example is the Soviet development of the supersonic transport. Now that the American Congress has proved itself strong enough to fend off the pressure of the American military industrial complex, it remains to be seen if conservationist forces in the USSR, France, and England will be as strong. Conceivably, the Russians would restrict their supersonic transport to the confines of their borders. In this way they might argue that the decision to fly it or not is purely an internal matter. It is more likely, however, that the Russian supersonic transport, the TU 144, like the French and British Concorde is intended for transatlantic and international

travel. One of the ports of call of the Russian plane
reportedly is to be India (*WSJ* 5/14/71, p. 1). Even if the
plane stays within Soviet borders, there are many who
feel that no matter where the supersonic transport flies,
the effects will be felt all over the planet. Dr. Harry
Johnston of the University of California at Berkeley
theorizes that operating five hundred supersonic trans-
ports an average of 7 hours a day might reduce the ozone
in the earth's atmosphere by 50 percent in six months to
a year. Destruction or depletion of the ozone belt would
significantly increase the amount of ultraviolet radiation
which reaches the surface of the earth. As a result almost
all vegetation would burn up except that which is under
water. Thus, regardless of whether one lived in the USSR
or outside of it there would be no escaping the effects.
These are issues that so far have been kept from the
Soviet public. As of late 1971, the supersonic transport
had been presented only as a symbol of pride and prowess, not of pollution and perversity. Little or no mention
has been made about its disastrous potential. Even
though there may be only one chance in one hundred
that Johnston is correct, such theories should be discussed
and all production halted until they are either proved or
disproved. This holds for the production and flight of
military as well as civilian supersonic aircraft. The
Russians as well as the Americans owe our planet that
much.

The quest of the USSR, England, and France for the
supersonic transport is perhaps one of the most frightening aspects of what the unimpeded introduction of tech-

nology can do. Fortunately, most of the countries of the
world were able to head off a threat of similar magnitude
which stemmed from the unrestrained testing of nuclear
weapons. But then the world had already seen a graphic
demonstration of what could happen when an atomic
bomb was detonated, and it was aware of the hazards
of the rising radioactivity count. The effect of such
"advances" as insecticides, defoliation, and the supersonic
transport is more problematic. Furthermore, there are
also issues of national prestige and technological exper-
tise involved. The question is, Can the world tolerate
such selfish and self-aggrandizing efforts by any country
in the future? As for the supersonic transport, failure to
curb the ecological masochism of the military industrial
complex in the USSR, France, and England only increases the pressure in the United States from our military industrial complex to resume development work on
our own supersonic transport and other potential ecological disasters. How long can we tolerate the operation
of the Concorde or TU 144 without an American
counterpart?

In a similar vein, lower environmental standards in one
country impose competitive burdens on countries with
higher standards. There is a danger that a Gresham's
law of environmental degradation will operate. Those
countries with poorer standards may tend to pull down
those with higher standards. Fortunately, this has not
happened in any massive way so far. Worldwide concern
for a quality environment has tended to force at least the
developed countries to direct increased attention to the

control of environmental disruption. Even in Japan enormous pressures have been generated despite the fact that this will make Japanese goods less competitive in foreign markets. But can pressure enough be built up to induce the Soviet government to respond, or is the system such that the government can divert or deflate such domestic pressure before it can exert enough weight? If the Soviet Union fails to respond to domestic pressure for better environmental quality, another possibility is international pressure. The Russians frequently respond more readily to international public opinion than their own. For that matter, international pressure should be applied to all countries that refuse to abide by international standards. Just as worldwide demonstrations were carried out against the United States because of its testing of atomic weapons and its defoliation program, so protests and demonstrations should be conducted against anyone who flies a supersonic transport or refuses to institute strict environmental controls. The consequences are just as grave.

There is a danger that poorer countries may view this as a plot by the richer countries of the world to force producers in emerging countries to assume social costs that their predecessors managed to shirk. There is an element of truth to such charges, but we may no longer be able to afford such political and economic pettiness. Regardless of the stage of economic development and regardless of the political system, the earth's environment is such a delicate mechanism and our technological powers are becoming so enormous that we have less and less choice in the matter. So far no one political system, be it the American model or Soviet socialism, seems to have a monopoly on all the right answers. Far from it. We must learn to cooperate with each other and our environment. If we fail to do so, no one will be the winner.

Appendix A Selected
Laws on the Environment

1910–1919

1914

Zemstvo and Moscow City Council set up Zone of Sanitary Protection for the Moskvorets Water Supply.
(Kolbasov 1965, p. 104)

Nov. 8, 1917

Nationalization of Land. (Kazantsev 1967, p. 6)

May 17, 1918

Sovnarkom Decree "On Organizing Irrigation Work in Turkestan." (Kolbasov 1965, p. 116)

May 27, 1918

"Decree About Forests" and supplementary laws in 1920 and 1923 which attempted to protect the forests, parks, and gardens from the ravages of the world and civil wars and the exploitation of the cold and hungry populace.
(Kazantsev 1967, p. 8; Zile 1970a, p. 11)

Dec. 1918

Fishing and Poaching Law supplemented in February 1919, February 26, 1920, May 25, 1922, and December 5, 1922. (Kazantsev 1967, pp. 9–10; Zile 1970a, pp. 17–20)

1919

Creation of Astrakhan National Forest Preserve (Kazantsev 1967, p. 11)

Feb. 14, 1919

Socialist Land Use regulating irrigation and drainage and the use of natural resources. (Kazantsev 1967, pp. 6–7)

Feb. 20, 1919

Supreme Council of National Economy adopted decree "On the Central Committee of Water Protection." (Kolbasov 1965, p. 212)

June 18, 1919

"The Sanitary Protection of Conditions in Residential Areas" to control sewage.

May 27, 1919, and July 20, 1920

Regulations for Hunting. (Kazantsev 1967, p. 10)

May 14, 1919

On Recording Water Wells. (Kolbasov 1965, p. 116)

1920–1929

1924

People's Commissariat on Health and Internal Affairs recommended that sanitary protection zones be set up. (Kolbasov 1965, p. 104)

Nov. 21, 1924

Sovnarkom Resolution on Fishing. (Kazantsev 1967, p. 10)

July 6, 1928

Council of People's Commissars of RSFSR, Ukraine, and Belorussia "On Establishment of Sanitary Protection of Water Sources." (Kolbasov 1965, p. 104)

1930–1939

May 17, 1937

Central Executive Committee and Sovnarkom of USSR "On the Sanitary Protection of Water Supply System and Sources of Water Supply." (Kolbasov 1965, pp. 104, 220)

1940–1949

April 10, 1940

Protection of Health Resorts (Kazantsev 1967, p. 50)

May 31, 1947

Council of Ministers of USSR "On Measures to Eliminate Pollution and to Provide the Sanitary Protection of Water Resources." (Kolbasov 1965, p. 214)

Oct. 1948

Creation of Shelter Belt of Trees. (Shabad 1951, p. 58)

1949

Council of Ministers of USSR, resolution for the control of air pollution and establishment of Chief Administration for Sanitary Epidemiological Supervision. (*LG* 8/9/67, p. 10)

1950–1959

Feb. 25, 1955

Regulation of Use of Timber by Kolkozniks. (Kazantsev 1967, p. 44)

Law for the Preservation of Nature
Estonia, June 7, 1957
Armenia, May 14, 1958
Georgia, Nov. 28, 1958
Lithuania, April 22, 1959
Moldavia, May 16, 1959
Azerbaijan, June 15, 1959
Uzbekistan, Nov. 19, 1959
Tadzhikistan, Nov. 25, 1959
Latvia, Nov. 27, 1959
Ukraine, June 30, 1960
RSFSR, Oct. 27, 1960
Belorussia, Dec. 21, 1961
Kazakhstan, May 12, 1962
Kirgizia, May 22, 1962
Turkmenia, May 26, 1963
(Kazantsev 1967, pp. 14–16, 37)

1958

Some Air Pollution Laws. (Sushkov 1969, p. 6)

Sept. 15, 1958

Fishing Laws. (Kazantsev 1967, p. 13)

May 11, 1959

Hunting Laws. (Kazantsev 1967, p. 13)

Sept. 4, 1959

Council of Ministers of USSR, "On Strengthening State Control over Use of Underground Water and Measures to Protect Them." (Kolbasov 1965, pp. 122, 234)

Oct. 15, 1959
Geological Work Regulations. (Kazantsev 1967, p. 14)

1960–1969
April 1960
The Ministry of Geological Work and Chief State Sanitary Inspector, Utilization and Protection of Underground Waters in the USSR. (Kolbasov 1965, p. 234)

April 22, 1960
On Measures to Regularize the Use and Increase the Protection of the Water Resources of the USSR. (Kolbasov 1965, p. 214)

May 9, 1960
Council of Ministers of the RSFSR "On the Preservation and Use of the Natural Wealth in the Basin of Lake Baikal." (Trofiuk and Gerasimov 1965, p. 58)

Aug. 9, 1960
RSFSR "Regulation for the Use of Communal Water Supplies and Sewage System." (Kolbasov 1965, p. 54)

Sept. 8, 1960
Control of Mining Wastes in RSFSR. (Kazantsev 1967, p. 14)

Mar. 3, 1962
Council of Ministers of RSFSR "Protection of Beaches on Black Sea in Krasnodarsk Krae." (Kazantsev 1967, p. 50)

Jan. 8, 1963
Special Law on Air Pollution for Moscow. (Kazantsev 1967, p. 55)

Feb. 18, 1963

Penalties for Hurting Nature and Water in RSFSR.
(Kazantsev 1967, p. 58)

Oct. 29, 1963

Council of Ministers of USSR defines duties for State
Sanitary Inspector.

Mar. 27, 1964

Law on Fishing and Fish Stocks. (Kazantsev 1967, p. 58)

Feb. 1, 1965

Moscow Green Belts. (Kazantsev 1967, p. 47)

Apr. 7, 1965

Green Belts, RSFSR. (Kazantsev 1967, p. 46)

Oct. 26, 1965

Laws on Air and Nature for RSFSR. (Kazantsev 1967,
p. 54)

June, 1966

Council of Ministers of Soviet Union ordered discontinuation of log floating and rafting. (*Prav* 8/31/69, p. 2)

Sept. 15, 1966

Regulation of Use of Timber by Kolkhozniks in RSFSR.
(Kazantsev 1967, p. 45)

Oct. 26, 1966

Measures for Denying Premium for Underfulfillment in
Established Period and Measures for Stopping Pollution
of Fish Ponds. (*Sov Lat* 2/8/70, p. 4)

Mar. 20, 1967
Law on Urgent Measures for the Protection of Soil from
Wind and Water Erosion. (*LG* 2/25/70, p. 12)

Oct. 2, 1968
Council of Ministers of USSR, "On Measures to Avert
the Pollution of the Caspian Sea." (*Prav* 10/3/68, p. 2)

Dec. 21, 1968
Preservation of Nature in Latvia. (*Sov Lat* 12/22/68,
pp. 3–4)

Feb. 7, 1969
Law on the Preservation of Lake Baikal. (*Iz* 2/8/69, p. 2)

Feb. 26, 1969
"On the Urgent Measures for Protecting the Black Sea
Coast from Destruction and for Rational Utilization of
the Resort Territory of the Black Sea Coast." (*Prav*
2/26/69, p. 3)

Dec. 23, 1969
New Law on Public Health—firms need an approved
plan of action on sanitation before they can open. (*Med
Gaz* 12/23/69)

1970
Dec. 12, 1970
Principles of Water Legislation of the USSR and the
Union Republics. (*CDSP* 1/26/71, p. 7)

1971
Sept. 24, 1971
Law on the Preservation of Lake Baikal.
(*Sot In* 9/24/71, p. 1)

Appendix B The Conservation Law of the Russian Republic (RSFSR) 1960

Russian Republic Law: ON CONSERVATION IN THE RUSSIAN REPUBLIC. (Pravda, Oct. 28, 1960, p. 2. Complete text:) Nature and its resources in the Soviet state comprise the natural basis for development of the national economy, serve as the source of the steady growth of material and cultural wealth and ensure the best conditions of work and rest for the people.

The Soviet social system and planned management of the economy create the possibility of rational use of the natural resources of the Russian Federation.

In the years of Soviet rule considerable work has been done in the Russian Republic in organizing the conservation and rational use of natural resources. But there are still substantial shortcomings in the matter of conservation.

In the period of the comprehensive building of communism the economic use of our country's rich natural resources is being intensified and the distribution of production forces throughout its territory is being greatly improved. This makes it necessary to establish a system of measures aimed at the protection, rational use and expanded reproduction of natural resources.

Conservation is a major state task and a concern of all the people.

Ministries and agencies should take into account the interests of allied

Reprinted from *Current Digest of the Soviet Press*, Vol. XII, No. 44, November 30, 1960. Translation copyright 1970 by The Current Digest of the Soviet Press, published weekly at the Ohio State University by the American Association for the Advancement of Slavic Studies; reprinted by permission.

branches of the economy and of the national economy as a whole, as well as the needs of the population, in solving national economic problems concerned with the development of new regions and the reconstruction of already developed ones, the reorganization of river systems, the shifting of vast areas to artificial irrigation and the use of individual natural resources.

With a view to strengthening conservation and ensuring the rational use and reproduction of natural resources, the Russian Republic Supreme Soviet *resolves that:*

ART. 1. NATURAL RESOURCES SUBJECT TO CONSERVATION. All the following natural resources, both in economic use and unexploited, are subject to state protection and regulated use in the territory of the Russian Republic:

(a) land;

(b) mineral resources;

(c) waters (surface and underground and soil moisture);

(d) forests and other natural vegetation and green plantings in population centers;

(e) typical landscapes and rare and remarkable natural objects;

(f) resort areas, forest-park shelter belts and natural green zones;

(g) the animal world (useful wild fauna);

(h) the atmosphere.

ART. 2. CONSERVATION OF LANDS. All lands, especially arable lands assigned to land users as the basic means of production in agriculture are subject to protection.

All land users are required to carry out systematically, with due consideration of local conditions, a complex of agrotechnical meliorative and anti-erosion measures aimed at preservation of the soil covering and maintenance of the most advantageous regimen of soil moisture and fertility.

All state and collective farms and other organizations are required to possess information on the nature and features of the soils of lands belonging to them for the purpose of rational organization of the application of fertilizer and correct control of processes that regulate the life of the soil itself and affect the harvest of crops grown on it.

A calculation of agricultural lands according to their economic worth and quality of the soil and compilation of a land cadastre is carried out by the Russian Republic Ministry of Agriculture.

The agricultural use of soils and the use of other natural resources connected with soils (vegetation and waters) must not result in a reduction of the agricultural land area or in a deterioration of the quality of fertile lands.

On lands subject to water and wind erosion, land users must carry out an obligatory complex of anti-erosion measures established with due consideration of local conditions.

Enterprises and organizations engaged in construction and prospecting work and the extraction of minerals (including building materials and peat) are required to carry out measures to restore the soil fertility of lands affected by their projects and suitable for agricultural use.

In agrotechnical and forest-exploitation work and in road, hydrotechnical and other types of construction, it is forbidden to use devices and methods that contribute to the development of water and wind erosion of soils (the washing away, blowing away and displacement of soils and ground, the development of ravines, the scattering of sands, and the formation of flood freshets and landslides), to salinization or bogging of soils and to other forms of loss of soil fertility.

The agricultural use of lands the exploitation of which could lead to the development of the above harmful processes is also forbidden.

ART. 3. CONSERVATION OF MINERAL RESOURCES. Reserves of solid, liquid and gaseous minerals found in the earth are subject to protection as a source of mineral raw material and fuel for the national economy; also subject to protection are classical and supporting geological outcrops that serve to determine the age of rocks and are of great importance to science and production.

Ministries, agencies and economic councils and enterprises under them that are engaged in the extraction of minerals are required, under the supervision of republic geology and conservation agencies, to ensure the safety of work and the exploitation of deposits in accordance with established norms and regulations and with due consideration of their fullest and most integrated use and economic expediency.

ART. 4. CONSERVATION OF WATERS. Surface and underground waters are subject to protection against depletion, pollution and obstruction and to regimen regulation as sources of water supply for the population and the national economy, sources of power, transportation routes, places of the growth of useful water vegetation, habitats of fish and water animals, hunting grounds, places of rest and tourism, curative resources and objects of interest to science, education and culture.

All organziations whose activity affects the water regimen are required:
(a) to carry out in the areas used hydromeliorative, forestmeliorative, agrotechnical and sanitary measures that will improve the water regimen and prevent the possibility of harmful effects on water (floods, heating, bogging, salinization, soil erosion, formation of ravines, freshets, etc.);
(b) to use water sources without exceeding established norms and to expend irrigation, ground and artesian waters prudently, without permitting nonproductive use of them; to avoid the formation of nonproductive shoal waters in the construction of reservoirs;
(c) to build purification installations for artificial or natural purifying at

all enterprises that discharge polluted waters into bodies of water;

(d) to prevent the pollution and silting of spawning grounds and the obstruction of passageways to them as a result of timber floating;

(e) in designing hydrotechnical projects, to provide for measures that will ensure the protection and reproduction of fish resources.

It is forbidden to put into operation enterprises, shops and installations that discharge sewage without carrying out measures that will ensure purifying of it.

ART. 5. CONSERVATION OF FORESTS. Forests are subject to protection and regulated use as sources of timber and other technical raw material and of food and fodder products, as the habitats of useful animals and plants and as a major portion of the geographic environment having importance for water protection, water regulation, soil protection, field protection, climate, health, culture and esthetics. The planning of forestry and lumbering should be based not only on fully satisfying the lumber needs of the national economy and the population but also on the need for forest conservation and reforestation. The industrial felling of timber should be concentrated in the main in densely forested areas.

All timber users are required to carry out a complex of forestry measures aimed at the rapid reforestation of cutting areas with valuable tree varieties and at the protection of forests against fires, arbitrary cutting, harmful insects and trampling by livestock and to clear cutting areas promptly.

All enterprises, institutions and citizens must strictly observe the fire-safety regulations in forests.

Local Soviet executive committees, lumbering organizations, collective and state farms and other land users are required to adopt measures to improve and increase forest resources, to establish forests in sparsely forested areas and to set up shelter belts and other protective plantings.

In designing and building new cities and large centers and reconstructing old ones, ministries, agencies, economic councils and local Soviet executive committees shall ensure the preservation of forests that could serve as green zones, as well as green plantings within populated points.

The following is forbidden:

(a) the felling of more timber than the quota for annual use established for the given enterprise;

(b) the cutting of forests (except for maintenance measures) that are important for soil protection, field protection, water protection and water regulation, the zones of which are established by the Russian Republic Council of Ministers, as well as forests along the shores of lakes and rivers and their tributaries that are spawning areas for valuable commercial fish;

(c) the use on slopes of methods of felling and initial hauling of timber that result in ravage of forest soils and destruction of new growth;

(d) the use in cedar plantings of cutting methods that do not ensure their natural reforestation;

(e) the arbitrary cutting of timber, arbitrary construction in a forest area and arbitrary reclassification of forest areas;

(f) the pasturing of livestock in shelter belts and restricted forest belts, in young growth and forest plantings, in parks, forest-parks, urban forests and forest zones around population centers, and in orchards.

ART. 6. CONSERVATION OF OTHER NATURAL VEGETATION. Besides forests, natural (wild) vegetation is subject to protection and regulated use as a fodder base for domestic and useful wild animals, a source of food products, medicinal and technical raw material and wild plant seed for sowing, a reserve of species for planting, a means of reinforcing soils, and an essential part of the geographical environment that influences climate and the water regimen and enriches the soil. Individual valuable, rare and disappearing species of plants are also subject to protection.

With a view to maintaining and increasing the productivity of natural vegetation and also improving it, the pasturing of livestock should be regulated and carried out without an overuse of pastures, with due consideration of the periods required for the development of herbage and the state of the soil, with uniform use of an entire pasturing area and restriction of pasturing on new grass after hay cutting. State and collective farms and other organizations must adopt measures to improve natural fodder lands assigned to them.

Procurers of raw material from wild plants are forbidden to use predatory methods that impede renewal of useful plants and cause destruction of the vegetative covering.

ART. 7. CONSERVATION OF GREEN PLANTINGS IN POPULATION CENTERS. Green plantings in all population centers, as well as in green zones around them and along roads, are subject to protection as having sanitary, protective and cultural-esthetic importance.

The cutting of green plantings (except for maintenance measures) or the transplanting of them to other areas is permitted only as an exception requiring the permission of local Soviet executive committees under a procedure established by the Russian Republic Council of Ministers.

ART. 8. CONSERVATION OF TYPICAL LANDSCAPES AND RARE AND REMARKABLE NATURAL OBJECTS. Typical landscapes and rare and remarkable objects of organic and inorganic nature are subject to protection as characteristic or unique examples of natural conditions of individual zones and physical-geographical regions having scientific, cultural-cognitive and health value.

Local Soviet executive committees are required, in the interests of the present and future generations, to ensure the preservation of examples of virgin nature and picturesque places; natural objects that are of historical-commemorative value; hiking and excursion sights and places for the relaxation and treatment of the working people; natural laboratories for the study of the course of processes taking place in nature; centers for the propagation and settlement of valuable animals in order to enrich hunting grounds; and individual species of rare and disappearing plants and animals.

ART. 9. STATE GAME SANCTUARIES AND PRESERVES. The protection of areas and objects of nature with due consideration of the importance may be carried out through the organization of:

(a) state game sanctuaries, the territory of which is permanently with-drawn from economic use for research and cultural-enlightenment purposes;

(b) game preserves, on the territory of which economic use of only a part of the natural objects is permitted, and only at certain seasons, for certain periods of time and only to the extent that this does not harm the objects being protected.

The regimen of state game sanctuaries and preserves is established both for sizeable areas and for small landmarks (groves, lakes, portions of valleys and sea coasts, etc.) and individual objects (waterfalls, caves, unique geological outcrops, rare or historical trees, etc.), which are accordingly declared to be protected landmarks or monuments of nature.

Prospecting on the territory of state game sanctuaries is permitted only within the limits of their plan for scientific research work.

Territories are declared to be state game sanctuaries or preserves or protected landmarks or monuments of nature with the regimen of game sanctuaries under a procedure established by the Russian Republic Council of Ministers.

ART. 10. CONSERVATION OF RESORT AREAS, FOREST-PARK SHELTER BELTS AND SUBURBAN GREEN ZONES. In areas for the relaxation and treatment of the working people (health resort localities, zones for the sanitary protection of health resorts, forest-park shelter belts and suburban green zones) the entire aggregate of natural conditions contributing to the curative and health value of the areas is protected.

In health resort areas the natural objects that determine the basic specialization of the health resorts (mineral waters, muds, beaches, pine forests, etc.) are also subject to protection.

The autonomous-republic Councils of Ministers and territory, province and city Soviet executive committees establish protected zones with the regimen of game sanctuaries along interesting tourist routes and in the most frequently visited places for relaxation of the working people.

The development and reclamation of the above-enumerated objects and territories is carried out only in accordance with general development plans.

Ministries, agencies, economic councils and local Soviet executive committees, in planning the development or reclamation of the above-named objects and territories, and design organizations, in designing such development or reclamation, must provide for the preservation and improvement of the complex of their natural conditions.

To this end it is necessary:

(a) to see to it that the sites and projects of new communal and transport construction are so located that the curative qualities and landscape merits of a locality are preserved;

(b) to provide for the systematic conduct of landscaping and meliorative measures, including preventive work, against landslides, torrents, avalanches and cave-ins, washing away of shores, destruction of beaches, and littering and contamination.

ART. 11. CONSERVATION OF THE ANIMAL WORLD. Useful wild animals, fowl, fish, etc., found in a state of natural freedom are subject to protection and regulated use as resources for the sport of hunting and for commercial hunting, whaling and fishing as destroyers of harmful animals and a food base for commercial and other useful animals, as objects for later domestication and fur farming, as a reserve of species for introducing new forms and improving the fertility of domestic animals, etc.

Rare and disappearing species of animals are also subject to protection against destruction and extinction.

In this connection it is necessary:

(a) strictly to observe established hunting and fishing regulations;

(b) to help improve conditions for the existence and reproduction of animals through the preservation and improvement of habitats and migration routes;

(c) to regulate the use of commercial reserves, ensuring the necessary density and reproduction for commercial purposes.

(d) to encourage the growth of useful fauna, without destroying other useful wild animals, fish, fowl, etc., in the process;

(e) to carry out measures to combat harmful animals—destroyers of forests and crops, carriers of infections, and poisonous, parasitic and other predators that harm the economy.

It is forbidden to destroy noncommercial wild animals if they do not harm the economy or public health.

ART. 12. SANITARY PROTECTION OF NATURE. The atmosphere, surface and subsoil waters, soils and ground are subject to sanitary protection.

Local Soviet executive committees, institutions, enterprises and organizations are required to carry out measures that will prevent contamination of the atmosphere, surface and subsoil waters, soils and ground and the littering of areas.

Household and industrial refuse and waste are subject to use in the national economy or to systematic removal and disposal.

In designing enterprises and installations connected with the use of natural resources, ministries, agencies and economic councils are required to work out and introduce technological processes that will ensure maximum processing of raw material and fuel and will not allow harmful waste to enter atmosphere, surface bodies of water, subsoil waters and the soil.

If it is not possible to introduce technological processes and forms of production organization that prevent the discarding of products into the atmosphere, waters and the soil, it is necessary to set up efficient purification, disposal and recovery installations.

The content of harmful substances in products discarded into the atmosphere, water and soil must not exceed the maximum concentrations established with due consideration of all economic interests and hygienic norms.

ART. 13. CALCULATING THE QUANTITY AND QUALITY OF NATURAL RESOURCES. Ministries, agencies and economic councils engaged in the use and reproduction of natural resources are required to organize and carry out a qualitative and quantitative calculation of them by compiling cadastres, bloodstock inventories, special maps, etc.

The Russian Republic Council of Ministers' Central Statistical Administration is charged with organizing and systematizing the calculation of natural resources made by the ministries, agencies, economic councils and local Soviet executive committees of the Russian Republic.

ART. 14. PLANNING THE USE OF NATURAL WEALTH (RESOURCES). In working out plans for development of the national economy, planning and economic agencies are required:

(a) to consider the mutual relation of resources enumerated in Art. 1 so that the exploitation of certain resources does not harm others;

(b) in using natural resources that are not self-replenishing, to provide not only for the full satisfaction of the country's current needs but also for the saving and renewal of these resources on the basis of expanded reproduction;

(c) to provide for and allocate funds and other material resources for work in the protection and reproduction of natural resources regularly and in a planned fashion;

(d) to prevent the reduction of areas of useful natural land (forests, meadows, bodies of water) unless lands, enterprises, transportation routes or population centers of greater value are established in their place;

(e) in designing and carrying out industrial, transport, communal and other types of construction to ensure maximum preservation of valuable natural objects.

The planning of rational and integrated use of natural resources as well as of measures ensuring their restoration on the basis of expanded reproduction throughout the Russian Republic is entrusted to the Russian Republic Council of Ministers' State Planning Commission and the All-Russian Economic Council.

ART. 15. SUPERVISION OF CONSERVATION. The Russian Republic Council of Ministers, autonomous-republic Councils of Ministers, territory, province, district, city, settlement and rural Soviet executive committees, ministries, agencies and economic councils shall ensure supervision over the observance by institutions, enterprises, organizations, collective and state farms and citizens of existing conservation laws and over fulfillment of measures for the preservation and restoration of natural resources.

ART. 16. PARTICIPATION OF PUBLIC ORGANIZATIONS IN CONSERVATION. Conservation is a concern of all the people. Public organizations (trade union, youth, scientific and others) and voluntary societies participate in it with the enlistment of the broad masses of workers, collective farmers and the intelligentsia.

The All-Russian Society for Promoting Conservation and the Landscaping of Population Centers exercises guidance over all public work in the field of conservation.

The Russian Republic State Planning Commission, the All-Russian Economic Council and ministries and agencies shall enlist the participation of the All-Russian Society for Promoting Conservation and the Landscaping of Population Centers in the consideration of plans of comprehensive measures for the use and transformation of nature and of designs of major projects that affect the preservation and reproduction of natural resources.

Public conservation inspection services are set up to help state agencies under local branches of the All-Russian Society for Promoting Conservation and the Landscaping of Population Centers; these coordinate their work with other public inspection services (hunting, fishing, etc.).

The posts of public conservation inspectors are honorary.

ART. 17. RESEARCH WORK ON PROBLEMS OF CONSERVATION. Research institutes and higher educational institutions shall include in their plans for scientific work topics in the field of conservation and carry out a systematic study of permissible norms for the use of natural resources and possible means of their reproduction.

ART. 18. TEACHING THE PRINCIPLES OF CONSERVATION IN EDUCATIONAL INSTITUTIONS. In order to instill in young people a prudent attitude toward natural resources and habits of correct use of natural resources, the teaching of the principles of conservation shall be included in school programs and corresponding sections in textbooks on the natural sciences, geography and chemistry; obligatory courses on the conservation and reproduction of natural resources shall be introduced in higher and specialized secondary educational institutions with due consideration of their specialization.

ART. 19. PROPAGANDIZING QUESTIONS OF CONSERVATION. Cultural-enlightenment institutions and organizations, publishing houses, museums, motion pictures, radio, television, newspapers and magazines, and voluntary societies shall widely propagandize tasks in the conservation and reproduction of natural resources.

ART. 20. LIABILITY OF DIRECTORS OF AGENCIES AND ENTERPRISES. Institutions, enterprises and organizations that have been assigned land and other natural resources for use or exploitation are required to ensure the protection, rational exploitation and reproduction of the natural resources.

For unlawful destruction or damage of natural resources the directors of institutions, enterprises and organizations as well as other persons directly guilty of committing such damage are held liable under the procedure established by law.

ART. 21. LIABILITY OF CITIZENS GUILTY OF UNLAWFUL USE OR DAMAGE OF NATURAL RESOURCES. Citizens who are guilty of the unlawful use or damage of natural resources are subject to administrative or criminal liability under the procedure established by law, with the recovery from them of losses caused.

ART. 22. THE DRAFTING AND IMPLEMENTATION OF CONSERVATION MEASURES. The Russian Republic Council of Ministers is charged with drafting and carrying out the necessary measures, stemming from this law, for conservation in the Russian Republic.

N. Organov,
Chairman of the Presidium, Russian Republic Supreme Soviet.
S. Orlov,
Secretary of the Presidium.

The Kremlin, Moscow, Oct. 27, 1960.

Appendix C Water Law, 1970

Laws and Resolutions Adopted by the U.S.S.R. Supreme Soviet:
PRINCIPLES OF WATER LEGISLATION OF THE U.S.S.R. AND
THE UNION REPUBLICS.* (Pravda, Dec. 12, pp. 2–3; Izvestia, Dec.
11. Complete text:) As a result of the victory of the Great October Socialist
Revolution, water, like all other natural resources in our country, *was
nationalized* and became the property of the people.

State ownership of water [which came about as the result of nationaliza-
tion] constitutes the basis of water relations in the U.S.S.R., creates favorable conditions for carrying out the [scientifically substantiated] planned
and integrated utilization of water with the greatest national-economic
effect and makes it possible to provide *Soviet* people with the best working,
living, recreation and public health conditions.

[Population growth and] The development of social production and
urban construction *and the rise in the population's material well-being and
cultural level* are increasing the all-round requirements for water and

* [The draft Principles of Water Legislation of the U.S.S.R. and the Union
Republics appeared in the Current Digest of the Soviet Press, Vol. XXII,
No. 17, pp. 10–14, 23. Changes in and additions to the draft version are
indicated in italics. Substantive passages dropped from the draft are en-
closed in brackets.]

Reprinted from *Current Digest of the Soviet Press*, Vol. XXII, No. 52, January
26, 1971. Translation copyright 1971 by The Current Digest of the Soviet
Press, published weekly at the Ohio State University by the American
Association for the Advancement of Slavic Studies; reprinted by permission.

heightening the importance of the rational utilization and conservation of water.

Soviet water legislation is called upon actively to facilitate the most effective and scientifically substantiated utilization of water and its protection against pollution, obstruction and depletion.

[These Principles define the goals of Soviet water legislation and establish general provisions concerning the procedure for the use of water, its conservation, prevention against the harmful action of water, and the organization of state record-keeping and planning of the utilization of water.]

I. General Provisions

ART. 1. GOALS OF SOVIET WATER LEGISLATION. The goals of Soviet water legislation are the regulation of water relations for the purpose of ensuring the rational utilization of water for the needs of the population and the national economy, protecting water from pollution, obstruction and depletion, preventing *and eliminating* the harmful action of water, improving the condition of bodies of water, protecting the rights of enterprises, organizations, institutions and citizens and strengthening legality in the field of water relations.

ART. 2. WATER LEGISLATION OF THE U.S.S.R. AND THE UNION REPUBLICS. Water relations in the U.S.S.R. are regulated by these Principles and by other acts of U.S.S.R. water legislation issued in accordance with these Principles, as well as by water codes and other acts of Union-republic water legislation.

Land, forest and mining relations are regulated by appropriate U.S.S.R. and Union-republic legislation.

ART. 3. STATE OWNERSHIP OF WATER IN THE U.S.S.R. In accordance with the U.S.S.R. Constitution, water in the Union of Soviet Socialist Republics is state property, i.e., the property of all the people.

Water in the U.S.S.R. is exclusively the property of the state, and only the right to the use of water is granted. Actions that infringe, directly or in a latent form, on the right of state ownership of water are prohibited.

ART. 4. THE UNIFIED STATE WATER SUPPLY. All water (bodies of water) in the U.S.S.R. constitutes the unified state water supply.

The unified state water supply consists of:

(1) rivers, lakes, reservoirs, other surface bodies of water and water sources, and the water of canals and ponds;

(2) *underground water and glaciers;* [the inland seas and territorial seas of the U.S.S.R.]

(3) *the inland seas and other inland sea waters of the U.S.S.R.* [underground water];

(4) *the territorial waters (territorial seas) of the U.S.S.R.* [glaciers].

ART. 5. THE COMPETENCE OF THE U.S.S.R. IN THE FIELD OF THE REGULATION OF WATER RELATIONS. The jurisdiction of the U.S.S.R. in the field of the regulation of water relations includes:

(1) the administration of the unified state water supply within the limits necessary to exercise the authority of the U.S.S.R. in accordance with the U.S.S.R. Constitution;

(2) the establishment of basic provisions in the field of the utilization of water, its protection from pollution, obstruction and depletion, and the prevention *and elimination* of the harmful action of water [as well as the planning of all-Union measures in this field];

(3) the establishment of all-Union normatives for water use and water quality and methods for *evaluating water quality* [determining these normatives];

(4) the establishment of a *unified system of state record-keeping of water and its utilization, the registration of water use and a* state water cadastre for the U.S.S.R. [and the procedure of its administration];

(5) the confirmation of schemes for the integrated utilization and conservation of water and also of [state] water-resource balances of all-Union importance;

(6) *the planning of all-Union measures for the utilization and conservation of water and for the prevention and elimination of the harmful action of water* [the organization of supervision over the utilization and conservation of water];

(7) *state supervision over the utilization and conservation of water and the establishment of procedures for the implementation of this supervision* [the regulation of the use of bodies of water situated on the territory of more than one Union republic];

(8) *the determination of bodies of water the regulation of the use of which is exercised by U.S.S.R. agencies.*

ART. 6. THE COMPETENCE OF THE UNION REPUBLICS IN THE FIELD OF THE REGULATION OF WATER RELATIONS. The jurisdiction of the Union republics in the field of the regulation of water relations, *outside the limits of the competence of the U.S.S.R.,* includes: the administration of the unified state water supply *on the republic's territory* [within the rights of the Union republics]; the establishment of procedures for the use of water, its protection from pollution, obstruction and depletion and the prevention and elimination of the harmful action of water; the planning of [republic] measures for the utilization and conservation of water and for the prevention and elimination of the harmful effects of water; the confirmation of schemes for the integrated utilization and conservation of water and of water-resource balances [of republic importance]; the exercise of *state* supervision over the utilization and conservation of water; and the regulation of water relations on other questions, providing they do not fall within the competence of the U.S.S.R.

ART. 7. STATE ADMINISTRATION IN THE FIELD OF THE UTILIZATION AND CON-
SERVATION OF WATER. State administration in the field of the utilization
and conservation of water is exercised by *the U.S.S.R. Council of Ministers,
the Union-republic Councils of Ministers, the autonomous-republic Councils of
Ministers and the executive committees of local Soviets, and also by specially empowered state agencies for the regulation of the utilization and conservation of water,
directly or through basin (territorial) administrations and other state agencies,
according to U.S.S.R. and Union-republic legislation.* [agencies for the regulation of the utilization and conservation of water, directly or through basin
(territorial) administrations, according to the procedure established by
U.S.S.R. legislation. The executive committees of the local Soviets exercise
state administration in the field of the utilization and conservation of
water in accordance with U.S.S.R. and Union-republic legislation now in
force.

[These same agencies are charged with state supervision over the utilization and conservation of water and over the observance of water legislation.]

ART. 8. STATE SUPERVISION OVER THE UTILIZATION AND CONSERVATION OF
WATER. *The goal of state supervision over the utilization and conservation of water
is to ensure the observance by all ministries and departments, all state, cooperative and
public enterprises, organizations and institutions and all citizens of the established
procedures for the use of water, the fulfillment of obligations for the conservation of
water, the prevention and elimination of the harmful action of water, the regulations for
the keeping of records on water, and other regulations established by water legislation.*

*State supervision over the utilization and conservation of water is exercised by the
Soviets and their executive and administrative agencies, and also by specially empowered state agencies, according to the procedure established by U.S.S.R. legislation.*

ART. 9. THE PARTICIPATION OF PUBLIC ORGANIZATIONS AND CITIZENS IN THE
IMPLEMENTATION OF MEASURES FOR THE RATIONAL UTILIZATION AND CONSER-
VATION OF WATER. *Trade unions, young people's organizations, conservation
societies, scientific societies and other public organizations, as well as citizens, give
assistance to state agencies in the implementation of measures for the rational utilization and conservation of water.*

*Public organizations take part in activity aimed at ensuring the rational utilization
and conservation of water, in accordance with their statutes (charters) and U.S.S.R.
and Union-republic legislation.*

ART. 10. THE SITING, DESIGNING, CONSTRUCTION AND PUTTING INTO OPERA-
TION OF ENTERPRISES, INSTALLATIONS AND OTHER FACILITIES AFFECTING THE
CONDITION OF WATER [Art. 8 in the draft Principles]. In the siting, designing, construction and putting into operation of *new and reconstructed* enterprises, installations and other facilities and in the introduction of new technological processes affecting the condition of water, the rational utilization

of water is to be ensured, with priority for the satisfaction of the needs of
the population for drinking water and water for everyday use. In this
connection, measures are envisaged to ensure the keeping of records on
water removed from bodies of water and returned thereto, the protection
of water from pollution, obstruction and depletion, the prevention of the
harmful action of water [and, where appropriate, to ensure], *the restriction
of the flooding of land to the necessary minimum* [the minimum necessary flooding of land], the protection of land from salinization, rising ground water
or desiccation, and also the preservation of favorable natural conditions
and landscapes.

[Losses caused to enterprises, organizations and institutions, as well as to
citizens, during the implementation of water-resources measures (hydraulic
engineering operations, etc.) are subject to reimbursement according to the
procedure established by the U.S.S.R. Council of Ministers.]*

In the siting, designing, construction and putting into operation of *new
and reconstructed* enterprises, installations and other facilities on fishing-
industry bodies of water, additional measures are to be carried out in good
time to ensure the protection of fish [reserves] and of other [useful] aquatic
fauna and flora and conditions for their reproduction.

The sites for construction of [the above-mentioned] enterprises, installations *and other facilities affecting the condition of water* are to be determined in
agreement with the agencies for the regulation of the utilization and conservation of water, *the executive committees of the local Soviets*, the agencies
exercising state sanitary supervision, the agencies for the protection of fish
reserves and other agencies in accordance with U.S.S.R. and Union-
republic legislation. Designs for the construction of the above-mentioned
enterprises, installations *and other facilities* are subject to coordination with
the agencies for the regulation of the utilization and conservation of water
and other agencies in cases defined by U.S.S.R. legislation and according
to the procedure stipulated therein.

It is forbidden to put into operation the following:
new and reconstructed enterprises, shops, aggregates and communal and
other facilities that are not provided with devices to prevent the pollution
and obstruction of water or the harmful action of water;
irrigation and watering systems, *reservoirs and canals* before the *measures*
[devices] provided by the designs to prevent *flooding, rising ground water,
and the* bogging up, salinization and erosion of soil [are ready or measures
to prevent these conditions] have been taken;
drainage systems, before water intakes and other installations are ready
according to confirmed designs;

* [See Art. 20.]

water-removal installations, without fish-protecting devices in accordance
with confirmed designs;

hydraulic engineering installations, *before* [without] devices for the [dis-
charge] passage of flood water *and fish are ready* in accordance with confirmed designs;

*drilled water wells that are not equipped with water-regulating devices and for which,
in appropriate cases, sanitary protection zones have not been established.*

The filling of reservoirs is prohibited before the measures envisaged in the
designs for preparation of the reservoir bed have been carried out.

ART. 11. PROCEDURE FOR THE PERFORMANCE OF WORK ON BODIES OF WATER
AND IN ADJACENT REGIONS (ZONES) [Art. 9 in the draft Principles]. *Construction*, [hydraulic engineering operations, the extraction of useful minerals
and aquatic plants], dredging and explosive operations, *the extraction of
useful minerals and aquatic plants*, the laying of cables, pipelines and other
utilities lines, the cutting of timber and [construction] drilling, agricultural
and other operations on bodies of water or in adjacent regions (zones)
affecting the condition of water may be performed only with the agreement of
the agencies for the regulation of the utilization and conservation of water,
the executive committees of local Soviets and other agencies in accordance with
U.S.S.R. and Union-republic legislation.

II. Water Use

ART. 12. WATER USERS [Art. 10 in the draft Principles]. *State, cooperative and
public* enterprises, organizations and institutions and citizens can be water
users in the U.S.S.R.

In cases stipulated by U.S.S.R. legislation, other organizations and persons may also be water users.

ART. 13. OBJECTS OF WATER USE [Art. 11 in the draft Principles]. Use is
granted of the bodies of water enumerated in Art. 4 of these Principles.

The [economic] utilization of bodies of water of special state importance
or of special scientific or cultural value can be prohibited partially or completely according to the procedure established by the U.S.S.R. Council of
Ministers and the Union-republic Councils of Ministers.

ART. 14. TYPES OF WATER USE [Art. 12 in the draft Principles]. The use of
bodies of water is granted with the provision that the requirements and
conditions stipulated by law for the satisfaction of the population's needs
for drinking water and water for everyday, therapeutic, resort, health-
improvement and other purposes and of agricultural, industrial, power-
engineering, transportation, fishing-industry and other state and public
needs [including the discharge of sewage] are to be observed. *The utilization of bodies of water for the discharge of sewage can be permitted only in cases*

stipulated by U.S.S.R. and Union-republic legislation and with the observance of the special requirements and conditions stipulated therein.

A distinction is made between general water use, which is exercised without the employment of installations or technical devices *affecting the condition of water,* and special water use, which is exercised with the employment of *such* installations or devices. *In certain cases, bodies of water can also be assigned for special water use without the employment of installations or technical devices when this use would affect the condition of water.*

The list of types of special water use is established by the agencies for the regulation of the utilization and conservation of water.

Bodies of water may be in joint or single use.

Enterprises, organizations and institutions that have been granted the single use of bodies of water are the primary water users, and in cases established by U.S.S.R. and Union-republic legislation, have the right to authorize secondary water use by other enterprises, organizations and institutions, as well as citizens, with the agreement of the agencies for the regulation of the utilization and conservation of water.

ART. 15. PROCEDURE AND CONDITIONS FOR GRANTING THE USE OF BODIES OF WATER [Art. 13 in the draft Principles]. The use of bodies of water is granted above all for the satisfaction of the population's needs for drinking water and water for everyday purposes.

The single use of bodies of water is granted fully or partially on the basis of resolutions of the Union-republic Councils of Ministers or the autonomous-republic Councils of Ministers or decisions of the executive committee of the appropriate Soviet or of another specially empowered state agency, according to the procedure established by U.S.S.R. and Union-republic legislation.

Special water use is exercised on the basis of authorizations issued by the agencies for the regulation of the utilization and conservation of water, *and, in cases stipulated by U.S.S.R. and Union-republic legislation, by the executive committees of local Soviets.* Such authorizations are issued after coordination with [the executive committees of the local Soviets] the agencies exercising state sanitary supervision and the agencies for the protection of fish reserves [(with respect to bodies of water used in the fishing industry), and, when necessary], and also with other interested agencies. The procedure for coordination and the issuance of authorizations for special water use is established by the U.S.S.R. Council of Ministers.

General water use is exercised without any authorization, according to the procedure established by Union-republic legislation. For bodies of water that have been granted for single use, general water use is permitted according to conditions established by the primary water user, with the agreement of the agencies for the regulation of the utilization and conservation of water, and when necessary may be prohibited.

Water use is exercised free of charge. Special water use may be subject to a fee, in cases and according to the procedure established by the U.S.S.R. Council of Ministers.

ART. 16. TIME PERIODS OF WATER USE [Art. 14 in the draft Principles]. Bodies of water are granted for indefinite or temporary use.

Water use without a time period set in advance is called indefinite (permanent).

Temporary use may be short-term—up to three years—or long-term—from three to 25 years. When necessary, the time periods for water use may be extended for a period not to exceed the upper limit of short-term or long-term temporary use, whichever is appropriate.

General water use is not limited in time.

ART. 17. RIGHTS AND DUTIES OF WATER USERS [Art. 15 in the draft Principles]. Water users have the right to use bodies of water only for those purposes for which use has been granted.

In cases stipulated by U.S.S.R. and Union-republic legislation, the rights of water users may be limited in the interests of the state, as well as in the interests of other water users. *At the same time, the conditions of the use of bodies of water for the population's needs for drinking water and water for everyday purposes must not deteriorate.* [The right to the use of water for the population's needs for drinking water for everyday purposes is not subject to limitation.]

Water users are obliged:

to utilize bodies of water rationally and to be concerned about the economical expenditure of water and about restoring and improving the quality of water;

to take steps for the complete cessation of the discharge into bodies of water of sewage containing pollutants;

[in the process of using water] not to violate the rights granted to other water users, and also not to damage economic facilities or natural resources (land, forests, *living things*, useful minerals, etc.);

to maintain in good condition purification and other water-resource installations and technical devices affecting the condition of water, to improve their operational qualities, and in prescribed cases to keep a record of water use.

ART. 18. GROUNDS FOR DISCONTINUING THE RIGHT TO WATER USE [Art. 16 in the draft Principles]. The rights of enterprises, organizations, institutions and citizens to the use of water are subject to discontinuation in the following instances:

(1) termination of the need for or renunciation of water use;

(2) expiration of the time period for water use;

(3) liquidation of the enterprise, organization or institution;

(4) transfer of water-resource installations to other water users;

(5) emergence of the necessity to remove a body of water from single use.

The rights of enterprises, organizations, institutions and citizens to water use (besides the right to use water for drinking and everyday needs) may be discontinued also in cases of the violation of the *regulations for the use of water and its conservation or in cases of the utilization of bodies of water for purposes not consonant with those for which it was granted* [regulations on the use and conservation of water].

Other grounds for discontinuing the rights of enterprises, organizations, institutions and citizens to water use may also be stipulated by Union-republic legislation.

ART. 19. PROCEDURE FOR DISCONTINUING THE RIGHT TO WATER USE [Art. 17 in the draft Principles]. The right to water use is discontinued through: abrogation of the authorization for special or secondary water use; withdrawal of bodies of water granted for single use.

The discontinuation of special water use is carried out at the decision of the agency that issued the authorization for this use.

Secondary water use may be discontinued at the decision of the primary water user, with the agreement of the agency for the regulation of the utilization and conservation of water.

The withdrawal of bodies of water from single use is carried out according to the procedure established by U.S.S.R. and Union-republic legislation.

The withdrawal of bodies of water from single use by enterprises, organizations and institutions of Union subordination *is* [may be] carried out on the basis of an agreement between the water users and the ministries and departments to which they are directly subordinate.

ART. 20. COMPENSATION FOR LOSSES CAUSED BY THE IMPLEMENTATION OF WATER-RESOURCES MEASURES OR BY THE DISCONTINUATION OF WATER USE OR CHANGES IN ITS CONDITIONS. *Losses caused to enterprises, organizations, institutions and citizens during the implementation of water-resources measures (hydraulic engineering operations, etc.), and also by the discontinuation of water use or changes in its conditions, are subject to reimbursement in cases stipulated by and according to the procedure established by the U.S.S.R. Council of Ministers.*

ART. 21. THE USE OF BODIES OF WATER FOR DRINKING, EVERYDAY AND OTHER NEEDS OF THE POPULATION [Art. 18 in the draft Principles]. Bodies of water in which the quality of the water corresponds to established sanitary requirements are granted for the drinking, everyday water-supply and other needs of the population.

The utilization of underground water of potable quality for needs not related to drinking and everyday water supply is, as a rule, not permitted. In areas where the necessary surface sources of water are lacking and where there are sufficient reserves of underground water of potable quality, the agencies for the regulation of the utilization and conservation of water may authorize the utilization of this water for purposes not related to drinking and everyday water supply [with the observance of the requirements established by Art. 36 of these Principles].

ART. 22. THE USE OF BODIES OF WATER FOR THERAPEUTIC, RESORT AND HEALTH-IMPROVEMENT PURPOSES [Art. 19 in the draft Principles]. Bodies of water that, according to established procedure, come within the category of therapeutic bodies of water are to be utilized primarily for therapeutic and resort purposes. In exceptional cases, the agencies for the regulation of the utilization and conservation of water may authorize the utilization of bodies of water coming within the therapeutic category for other purposes, with the agreement of the appropriate public health and resort-administration agencies [except for the utilization of these bodies of water for discharging sewage].

The discharge of sewage into bodies of water coming within the therapeutic category is prohibited.

The procedure for the use of water for recreation and sports is established by *U.S.S.R. and* Union-republic legislation.

ART. 23. THE USE OF BODIES OF WATER FOR THE NEEDS OF AGRICULTURE [Art. 20 in the draft Principles]. The use of bodies of water for the needs of agriculture is exercised according to the procedure for both general and special water use.

Special water use applies to the irrigation, water-supply, drainage [systems, wells] and other water-resource installations and devices belonging to [specialized] state organizations, collective farms, state farms and other water users.

Collective farms, state farms and other enterprises, organizations, institutions and citizens using bodies of water for agricultural needs are obliged to observe the established plans, regulations, norms and regimen of water use, to take steps to curtail losses of water through seepage and evaporation in *reclamation* [irrigation and water-supply] systems, *to prevent the irrational discharge of water into these systems, to prohibit the entry of fish into reclamation systems from fishing-industry bodies of water,* and to create the most favorable regimen of soil moisture. [In implementing agrotechnical, land-reclamation and other measures to ensure that the condition of the land is favorable to agricultural production, the above-mentioned water users are obliged not to cause any deterioration in the condition of bodies of water.]

The irrigation of agricultural land with sewage is authorized by the agencies for the regulation of the utilization and conservation of water, with the agreement of the agencies exercising state sanitary *and veterinary* supervision.

The provisions of this article extend also to the irrigation and drainage of land occupied by forests, forest belts and forest nurseries.

ART. 24. THE USE OF BODIES OF WATER FOR INDUSTRIAL PURPOSES [Art. 21 in the draft Principles]. Water users using bodies of water for industrial purposes are obliged to observe the established plans, technological norms and regulations for water use, and also to take steps to curtail the expenditure of water and to discontinue the discharge of sewage, through improving production technology and water-supply schemes (the use of water-free technological processes, air cooling, recycleable water supply and other technical methods).

The executive committees of the local Soviets, *in cases of natural disasters, accidents or other extraordinary circumstances, and also in cases in which enterprises exceed the established limits for the consumption of water from water lines, have the right to limit or prohibit the consumption for industrial purposes of drinking water from communal water lines and temporarily to restrict the consumption for industrial purposes of drinking water from departmental economic and drinking-water water lines in the interests of the priority satisfaction of the population's needs for drinking water and water for everyday purposes.*

Underground water (fresh, mineral and thermal) not coming within the category of drinking or therapeutic water may be utilized, according to established procedure, for technical water supply, the extraction of chemical elements contained in this water, the generation of thermal power and other production needs, with the observance of the requirements of the rational utilization and conservation of water.

ART. 25. THE USE OF BODIES OF WATER FOR THE NEEDS OF HYDRAULIC POWER ENGINEERING [Art. 22 in the draft Principles]. The use of bodies of water for the needs of hydraulic power engineering is exercised with consideration for the interests of other branches of the national economy, and also with the observance of the requirements of the integrated utilization of water, if no other direct provision has been made by a resolution of the U.S.S.R. Council of Ministers or a resolution of a Union-republic Council of Ministers or, in appropriate instances, by a decision of an agency for the regulation of the utilization and conservation of water.

ART. 26. THE USE OF BODIES OF WATER FOR THE NEEDS OF WATER TRANSPORTATION AND TIMBER FLOATING [Art. 23 in the draft Principles]. The rivers, lakes, reservoirs, canals, inland seas [and territorial seas] *and other inland sea waters of* the U.S.S.R., *and also the territorial waters (territorial seas) of the U.S.S.R.,* are waterways of general use, except in cases in which their utilization for these purposes is fully or partially prohibited or they have been granted for single use.

The procedure for assigning waterways to the navigation and timber-floating categories, as well as for establishing regulations for the operation of waterways, is determined by U.S.S.R. and Union-republic legislation.

The loose floating of timber, *and also the floating of timber in bundles and baskets not pulled by ships*, is prohibited:

(1) on navigable waterways;

(2) on bodies of water the list of which is confirmed by the U.S.S.R. Council of Ministers or by the Union-republic Councils of Ministers, in consideration of the special importance of these bodies of water for the fishing industry, water supply or other national-economic purposes.

On other bodies of water, the above-mentioned types of timber floating are permitted on the basis of authorizations issued by the agencies for the regulation of the utilization and conservation of water and after obtaining the agreement of the agencies exercising the protection of fish reserves [(3) on the bodies of water, without authorization from the agencies for the regulation of the utilization and conservation of water and the agreement of the agencies exercising the protection of fish reserves].

Timber-floating organizations are obliged to conduct the regular cleaning of sunken logs from waterways used for timber floating.

ART. 27. THE USE OF BODIES OF WATER FOR THE NEEDS OF AIR TRANSPORTATION [Art. 24 in the draft Principles]. The procedure for the use of bodies of water for the stationing, takeoff and landing of aircraft, as well as for other needs of air transportation, is established by U.S.S.R. legislation.

ART. 28. THE USE OF BODIES OF WATER FOR THE NEEDS OF THE FISHING INDUSTRY [Art. 25 in the draft Principles]. [The rivers, lakes, reservoirs, inland seas and other bodies of water that are utilized for the catching and breeding of fish or other aquatic fauna and flora or are of importance for the reproduction of fish reserves are recognized as fishing-industry bodies of water.].

On fishing-industry bodies of water or on individual sections of these bodies of water that are of especially great importance for the conservation and reproduction of valuable types of fish or other objects of water industries, the rights of water users may be limited in the interests of the fishing industry. The lists of such bodies of water or sections of bodies of water and the types of limitations on water use are determined by the agencies for the regulation of the utilization and conservation of water on the basis of presentations by the agencies exercising the protection of fish reserves.

In the operation of hydraulic-engineering and other installations on fishing-industry bodies of water, measures ensuring the protection of fish reserves and conditions for their reproduction must be carried out in good time.

The procedure for the use of bodies of water for the needs of the fishing industry is established by U.S.S.R. and Union-republic legislation.

ART. 29. THE USE OF BODIES OF WATER FOR THE NEEDS OF THE HUNTING INDUSTRY [Art. 26 in the draft Principles]. On rivers, lakes and other bodies of water inhabited by wild waterfowl and valuable fur-bearing animals (beaver, muskrat, desmans, coypu, etc.), the agencies for the regulation of the utilization and conservation of water may grant preferential water-use rights to hunting-industry enterprises and organizations, with consideration for the requirements of the integrated utilization of water.

The procedure for the use of bodies of water for the needs of the hunting industry is established by U.S.S.R. and Union-republic legislation.

ART. 30. THE USE OF BODIES OF WATER FOR THE NEEDS OF PRESERVES [Art. 27 in the draft Principles]. Bodies of water of particular scientific or cultural value are declared preserves, according to the procedure established by U.S.S.R. and Union-republic legislation, and are granted for indefinite single use as preserves, for purposes of conservation and the conduct of scientific research.

The procedure for the use of the water of preserves is determined by the statute on preserves.

The withdrawal of bodies of water from use as preserves is permitted only in cases of special necessity, on the basis of a resolution of a Union-republic Council of Ministers.

ART. 31. THE USE OF BODIES OF WATER FOR THE DISCHARGE OF SEWAGE [Art. 28 in the draft Principles]. The use of [rivers, lakes, seas and other] bodies of water for the discharge of industrial, communal, household, drainage and other sewage may be carried out only with the authorization of the agencies for the regulation of the utilization and conservation of water, after obtaining the agreement of the agencies exercising state sanitary supervision and the protection of fish reserves and of other interested agencies [and with the observance of the requirements stipulated in Art. 35 of these Principles].

The discharge of sewage is permitted only in cases in which it does not lead to an increase in the pollutant content of the particular body of water above established norms and on the condition that the water users purify the sewage up to the limits established by the agencies for the regulation of the utilization and conservation of water.

If the above requirements are violated, the discharge of sewage is to be limited, halted or prohibited by the agencies for the regulation of the utilization and conservation of water, up to and including discontinuation of the activity of individual industrial installations, shops, enterprises, organizations and institutions. In cases threatening the health of the population, the agencies exercising state sanitary supervision have the right to halt the discharge of sewage, up to and including discontinuation of the operation of production or other facilities, with notification to the agencies for the regulation of the utilization and conservation of water.*

The procedure and conditions for the use of bodies of water for the discharge of sewage are established by U.S.S.R. and Union-republic legislation.

ART. 32. THE USE OF BODIES OF WATER FOR FIRE-FIGHTING NEEDS AND OTHER STATE AND PUBLIC REQUIREMENTS [Art. 29 in the draft Principles]. The removal of water for fire-fighting needs is permitted from any body of water.

The procedure for the use of bodies of water for fire-fighting needs, as well as for other state and public requirements, is established by U.S.S.R. and Union-republic legislation.

ART. 33. THE OPERATION OF RESERVOIRS [Art. 30 in the draft Principles]. Enterprises, organizations and institutions that operate water-raising, water-transmission or water-removal installations on reservoirs are obliged to observe the regimen for the filling and operation of reservoirs that has been established, with consideration for the interests of water users and land users, in the zones affected by the reservoirs.

The procedure for the operation of reservoirs is determined by regulations confirmed by the agencies for the regulation of the utilization and conservation of water for each reservoir, cascade *or system* of reservoirs, with the agreement of the agencies exercising state sanitary supervision and the protection of fish reserves and of other interested agencies.

The organization and coordination of measures to ensure the proper technical condition and comprehensive outfitting of reservoirs and supervision over the observance [by water users of the requirements of water legislation and] of the regulations for reservoir operation is carried out by *the agencies for the regulation of the utilization and conservation of water* [specially empowered agencies], according to the procedure established by the U.S.S.R. Council of Ministers or the Union-republic Councils of Ministers.

The provisions of this article extend also to the operation of lakes and other bodies of water used as reservoirs.

ART. 34. REGULATION OF THE USE OF BODIES OF WATER SITUATED ON THE TERRITORY OF MORE THAN ONE UNION REPUBLIC [Art. 31 in the draft Principles]. The regulation of the use of bodies of water situated on the territory of two or more Union republics, in areas affecting the interests of these republics, is exercised by agreement between the agencies of the interested republics, with the exception of bodies of water for which the regulation of their use has been assigned to the competence of the U.S.S.R.

* [See bracketed paragraphs under Art. 38.]

[In the use of bodies of water situated on the territory of two or more Union republics, the water users of each republic are obliged not to violate the rights and interests of the water users of the other republic or republics.]

ART. 35. THE PROCEDURE FOR RESOLVING DISPUTES CONCERNING WATER USE [Art. 32 in the draft Principles]. Disputes concerning water use are resolved *by the Union-republic Councils of Ministers, the autonomous-republic Councils of Ministers, the executive committees of local Soviets, the agencies for the regulation of the utilization and conservation of water and other specially empowered state agencies,* according to [administrative] procedures established by U.S.S.R. and Union-republic legislation.

Disputes between water users of one Union republic and water users of another Union republic concerning water use are examined by a commission, set up on the basis of parity, of representatives of the interested Union republics. In cases in which the commission does not reach an agreed-upon decision, disputes concerning such questions are subject to examination according to a procedure determined by the U.S.S.R. Council of Ministers.

Property disputes involving *water relations* [the violation of rights to the use of water] are resolved by *procedures established by U.S.S.R. and Union-republic legislation* [arbitration or by legal procedure].

ART. 36. THE USE OF THE BORDER WATERS OF THE U.S.S.R. [Art. 33 in the draft Principles]. The use of U.S.S.R. border waters is exercised on the basis of international agreements.

To the extent that the use of the Soviet section of border waters is not regulated by international agreements to which the U.S.S.R. is a party, it is exercised in accordance with U.S.S.R. and Union-republic legislation.

The procedure for the use of U.S.S.R. border waters is established by the competent agencies, in coordination with the border troops command.

III. The Conservation of Water and the Prevention of its Harmful Action

ART. 37. THE CONSERVATION OF WATER [Art. 34 in the draft Principles]. All water *(bodies of water)* is subject to protection from pollution, *obstruction and depletion that may harm* [the results of which may entail a deterioration in] the health of the population *or entail* the diminution of fish reserves, *the deterioration of water-supply conditions* or other unfavorable phenomena as a consequence of changes in the physical, chemical or biological properties of the water, *a reduction in* [or] its capacity for natural purification *or the violation of the hydrological or hydrogeological regimen of water.*

[All water is also subject to protection from obstruction, depletion and other actions deleterious to its condition.]

Enterprises, organizations and institutions whose activity affects the condition of water are obliged to conduct—with the agreement of the agencies for the regulation of the utilization and conservation of water, *of the executive committees of local Soviets*, of the agencies exercising state sanitary supervision and the protection of fish reserves and of other interested state agencies *or at the direction of specially empowered state agencies* [and also with the agreement of the executive committees of local Soviets or at their direction]—technological, forest-amelioration, agrotechnical, hydraulic-engineering, sanitary and other measures to ensure the protection of water from pollution, obstruction and depletion, and also to improve the water's condition and regimen.

Measures for the conservation of water [that is of all-Union or republic importance] are stipulated in the state plans for the development of the national economy [of the U.S.S.R. and the Union republics].

ART. 38. THE PROTECTION OF WATER FROM POLLUTION AND OBSTRUCTION [Art. 35 in the draft Principles]. *The discharge of production, household and other types of waste materials into bodies of water is prohibited. The discharge of sewage is permitted only with the observance of the requirements stipulated in Art. 31 of these Principles.* [The utilization of bodies of water for the discharge of production, household and other types of waste materials, except sewage, is prohibited.

[The discharge of sewage is permitted only in cases in which it does not lead to an increase in the pollutant content of the particular body of water above established norms and on the condition that the water users purify the sewage up to the limits established by the agencies for the regulation of the utilization and conservation of water.

[If the above requirements are violated, the discharge of sewage is to be limited, halted or prohibited by the agencies for the regulation of the utilization and conservation of water, up to and including discontinuation of the activity of individual industrial installations, shops, enterprises, organizations or institutions. In cases threatening the health of the population, the agencies exercising state sanitary supervision have the right to halt the discharge of sewage, up to and including discontinuation of the operation of production or other facilities, with notification to the agencies for the regulation of the utilization and conservation of water.]

The owners of means of water transportation, pipelines [installations built on piles] and floating and other installations on bodies of water, timber-floating organizations *and other enterprises, organizations and institutions* are obliged not to permit the pollution or obstruction of water *as a consequence of losses of oils, wood, chemical, petroleum and other products* [by petroleum products, oils, wood and other types of waste materials].

Enterprises, organizations and institutions are obliged not to permit the pollution or obstruction of the surface of drainage collection systems, the ice cover of bodies or water or the surface of glaciers by industrial, everyday and other waste materials *and effluents* or by petroleum and chemical products whose contact with water entails a deterioration in the quality of surface or underground water.

The administrations of state water-resource systems, collective farms, state farms and other enterprises, organizations and institutions are obliged to prevent the pollution of water by fertilizers and toxic chemicals.

For the purpose of protecting water that is used for drinking and everyday water supply and for the therapeutic, resort and health-improvement needs of the population, regions and zones of sanitary protection are established in accordance with U.S.S.R. and Union-republic legislation.

ART. 39. THE PROTECTION OF WATER FROM DEPLETION [Art. 36 in the draft Principles]. To maintain a favorable water regimen in rivers, lakes, *reservoirs, underground water* [seas] and other bodies of water, [particularly] to prevent the water erosion of soil, the silting up of bodies of water *and the deterioration of habitat conditions for aquatic animals*, to reduce fluctuations in runoff water, etc., forest water-conservation zones are established and forest-amelioration, *anti-erosion, hydraulic-engineering* and other measures are conducted, according to the procedure stipulated by U.S.S.R. and Union-republic legislation.

The agencies for the regulation of the utilization and conservation of water, in coordinating questions of the siting and construction of enterprises, installations and other [economic] facilities affecting the condition of water, and also in issuing authorizations for special water use, must be guided by the schemes for the integrated utilization and conservation of water and by the [state] water-resource balances, which take into consideration the interests of the water users and land users [within the respective river, lake or inland-sea basin].

If, during the performance of drilling or other mining operations connected with the prospecting, surveying and exploitation of deposits of gas, petroleum, coal or other useful minerals, underground water tables are discovered, the organizations conducting the mining operations are obliged immediately to report this to the agencies for the regulation of the utilization and conservation of water and to take steps to protect the underground water, according to established procedure.

[The utilization of underground water of potable quality for needs not connected with the drinking and everyday water supply of the population is authorized by the agencies for the regulation of the utilization and conservation of water, in cases in which there is an especially good reason for doing so.]

Self-discharging wells are to be equipped with regulating devices, closed

down or dismantled, according to the procedure established by *U.S.S.R.
and* Union-republic legislation.

ART. 40. THE PREVENTION AND ELIMINATION OF THE HARMFUL ACTION OF
WATER [Art. 37 in the draft Principles]. Enterprises, organizations and
institutions are obliged to conduct, in coordination with the agencies for
the regulation of the utilization and conservation of water, the executive
committees of local Soviets and other *interested* state agencies or at *the
direction of specially empowered state agencies* [their direction], measures to
prevent and eliminate the following harmful actions of water:
floods, inundations and rising ground water;
the destruction of banks, protective dikes and other installations;
the bogging up and salinization of land;
the erosion of soil, the formation of ravines, landslides, flash floods and
other harmful phenomena.

The implementation of urgent measures to prevent and eliminate natural
disasters caused by the harmful action of water is regulated by U.S.S.R.
and Union-republic legislation.

Measures for the prevention and elimination of the harmful action of
water [that are of all-Union or republic importance] are provided for in
the state plans for the development of the national economy [of the
U.S.S.R. and the Union republics].

IV. State Record-keeping [Control] and Planning of the Utilization of Water

ART. 41. THE GOALS OF STATE RECORD-KEEPING AND PLANNING OF THE
UTILIZATION OF WATER [Organization of the Keeping of Records on Water
and of Control Over Its Utilization. Art. 38 in the draft Principles]. *The
state keeping of records on water and its utilization has the goals of establishing the
quantity and quality of water and of providing data on the utilization of water for
the needs of the population and the national economy.*

*The planning of the utilization of water must ensure the scientifically substantiated
distribution of water among water users, taking into consideration the priority satisfaction of the population's needs for drinking water and water for everyday purposes,
the conservation of water and the prevention of its harmful action. The planning of
the utilization of water must take into consideration the data of the state water
cadastre, the water-resources balances and the schemes for the integrated utilization
and conservation of water.*

[For the purpose of ensuring the rational and planned utilization and
conservation of water and preventing its harmful action, a state water
cadastre is kept, current and long-range state water-resource balances are
compiled, schemes for the integrated utilization and conservation of water
are worked out and state control is exercised.

[The goal of state control over the utilization of water is to ensure the observance by ministries and departments, by state, cooperative and public enterprises, organizations and institutions and by citizens of water legislation and the procedures for the utilization and conservation of water and for the prevention of its harmful action.

[State control over the utilization of water is carried out by specially empowered state agencies for the regulation of the utilization and conservation of water and by the executive committees of the local Soviets, according to the procedure established by U.S.S.R. legislation.]

ART. 42. THE STATE WATER CADASTRE [Art. 39 in the draft Principles]. The state water cadastre includes record-keeping data on water according to quantitative and qualitative indices, the registration of water use, and record-keeping data on the utilization of water.

ART. 43. THE [STATE] WATER-RESOURCE BALANCES [Art. 40 in the draft Principles]. [State] Water-resource balances, which evaluate the presence of water and the degree of its utilization, are compiled by basins, economic regions, Union republics *and the U.S.S.R.* [and are taken into consideration in the current and long-range planning of the development of the U.S.S.R. national-economy].

ART. 44. SCHEMES FOR THE INTEGRATED UTILIZATION AND CONSERVATION OF WATER [Art. 41 in the draft Principles]. General and basin (territorial) schemes for the integrated utilization and conservation of water determine the basic water-resource and other measures to be carried out for the satisfaction of the long-range requirements for water of the population and the national economy and for the conservation of water and the prevention of its harmful action.

ART. 45. THE PROCEDURE FOR THE KEEPING OF STATE RECORDS ON WATER AND ITS UTILIZATION AND KEEPING THE STATE WATER CADASTRE, FOR COMPILING [STATE] WATER-RESOURCE BALANCES AND FOR WORKING OUT SCHEMES FOR THE INTEGRATED UTILIZATION AND CONSERVATION OF WATER [Art. 42 in the draft Principles]. *The keeping of state records on water and its utilization,* the keeping of the state water cadastre, the compilation of [state] water-resource balances and the working out of schemes for the integrated utilization and conservation of water are carried out by the state, according to a single system for the U.S.S.R.

The procedure *for keeping state records on water and its utilization,* keeping the state water cadastre, compiling the [state] water-resource balances and working out and confirming the schemes for the integrated utilization and conservation of water is established by the U.S.S.R. Council of Ministers.

V. Liability for the Violation of Water Legislation

ART. 46. LIABILITY FOR THE VIOLATION OF WATER LEGISLATION [Art. 43 in the draft Principles]. [The unauthorized disposition of water] The ceding of the right to water use and other transactions that violate, overtly or covertly, the right of state ownership of water are invalid.

Persons guilty of the above transactions or of:

the unauthorized seizure of bodies of water or the unauthorized use of water;

the removal of water in violation of water-use plans;

the pollution *or obstruction* of water;

the putting into operation of enterprises, communal and other facilities without installations and devices to prevent *the* [water] pollution *and obstruction of water or its harmful action*;

the negligent utilization of water (extracted or drained from bodies of water);

the violation of the water-protection regimen in drainage collection systems, leading to their pollution, the water erosion of the soil and other harmful phenomena;

the unauthorized conduct of hydraulic-engineering operations;

the damaging of water-resource installations and devices;

the violation of regulations for the operation of water-resource installations and devices,

are criminally or administratively liable in accordance with U.S.S.R. and Union-republic legislation.

Union-republic legislation may establish liability for other types of violations of water legislation as well.

Bodies of water seized without authorization are to be returned to their authorized users, with no compensation for expenditures incurred during the unlawful use.

[ART. 44. COMPENSATION FOR LOSSES CAUSED BY THE VIOLATION OF WATER LEGISLATION.] Enterprises, organizations, institutions and citizens are obliged to make reimbursement for losses caused by the violation of water legislation, in amounts and according to the procedure established by U.S.S.R. and Union-republic legislation. *Officials and other employees through whose fault enterprises, organizations and institutions incur expenditure connected with reimbursement for losses are materially liable according to established procedure.*

Appendix D Conversion Table

1 ruble = $1.11*	$1 = 0.901 rubles
1 meter = 1.0936 yards	1 yard = 0.9144 meters
1 kilometer = 0.6214 miles	1 mile = 1.6093 kilometers
1 square meter = 10.764 square feet	1 square yard = 0.836 square meters
1 hectare = 2.47 acres	1 acre = 0.4047 hectares
1 square kilometer = 247.1 acres (0.386 square miles)	1 square mile = 2.59 square kilometers
1 liter = 0.264 gallons (U.S.)	1 gallon = 3.785 liters
1 kilogram = 2.205 pounds	1 pound = 0.4536 kilograms
1 centner = 100 kilograms = 220.5 pounds	1 pound = 0.004536 centners
1 cubic meter = 1.308 cubic yards	1 cubic yard = 0.7645 cubic meters
1 cubic kilometer = 0.24 cubic miles	1 cubic mile = 4.167 cubic kilometers

* In January 1972, the Russians revalued the ruble so that the prevailing rate after that period is

1 ruble = $1.20 $1 = 0.83 rubles.

All conversions in the book however are made at the rate of 1 ruble = $1.11.

Bibliography

Books and Articles
Abramov, B. S. 1963. "Skrytye istochniki zhizni" [The hidden sources of life]. *Priroda*, July.

Abramov, N. N. 1967. "Osnovnye dostizheniia sovetskoi vodoprovodnoi tekhniki i nauchnykh issledovanii v oblasti vodosnabzheniia za 50 let" [Fifty years of basic achievements of Soviet water distribution technology and scientific research in the field of water supply]. *Vodosnabzhenie i sanitarnaia tekhnika*, November.

————. 1970. "Perspektivy razvitiia i dal'neishego progressa v oblasti vodosnabzheniia" [Perspectives of development and further progress in the field of water supply]. *Vodosnabzhenie i sanitarnaia tekhnika*, April.

Akimovich, N. N., and Ramenskii, L. A. 1966. "Usloviia pogody i zagriaznennost' vozdukha v Odesse" [Weather conditions and air pollution in Odessa]. *Izvestiia akademii nauk, seriia geografiia*, No. 5.

Allakhverdian, D. 1968. "Ekonomicheskaia reforma i
voprosy khozrascheta" [Economic reform and questions
of economic accounting]. *Voprosy ekonomiki*, November,
pp. 3–15.

Allen, Gary. 1970. "Ecology: Governmental Control of
the Environment." *American Opinion*, May.

Apollov, B. A. 1962. "Znachenie ekonomicheskikh nauk
v reshenii 'Problemy Kaspiiskogo moria'" [The significance of economics in solving the "Problem of the Caspian Sea"]. *Voprosy geografii*, No. 57.

Apollov, B. A., and Bobrov, S. N. 1963. "Kaspiiskoe
more budet zhit'" [The Caspian Sea will live]. *Priroda*,
February.

Apollov, B. A., Giul', K. K., and Zavriev, V. G., eds.
1963. *Materialy Vsesoiuznogo soveshchaniia po probleme
Kaspiiskogo moria* [Materials of the All Union conference
on the problems of the Caspian Sea]. Baku: Izdatel'stvo
akademii nauk Azerbaidzhanskoi SSR, 1963.

Armand, D. 1966. *Nam i vnukam* [For us and our grand-
children]. Moscow: Mysl'.

Astrakhantsev, V. I. 1969. "Zashchitit' vodnye resursy
Irkutskoi oblasti ot zagriazneniia" [The protection of
water resources of the Irkutsk Oblast from pollution].
Gidrotekhnika i melioratsiia, January.

Astrakhantsev, V. I., and Pisarskii, B. I. 1968. "Vodok-
hoziaistvennye problemy pribaikal'ia" [Water management problems in the Baikal Region]. *Gidrotekhnika i
melioratsiia*, July.

Balandina, V. A. 1969. "Vliianie gazovykh vybrosov
kompleksa promyshlennykh predpriiatii neftekhimii na

zdorov'e detei doshkol'nogo vozrasta" [The influence of gaseous emissions of complex enterprises of the petro-chemical industry on the health of children of pre-school age]. *Nauchnye trudy Omskii meditsinskii institut*, No. 95.

Bannikov, A. G. 1968. "Ot zapovednika do prirodnogo parka" [From preserve to natural park]. *Priroda*, April.

Bazenkov, F. A. "Sud'ba vody v nashikh rehakh" [The fate of water in our rivers]. *Geografiia v shkole*, No. 5, 1968.

Belichenko, Iu. P. 1968. "Vsesoiuznoe soveshchanie po ispol'zovaniiu i okhrane vodnykh resursov" [The All Union conference on the use and preservation of water resources]. *Gidrotekhnika i melioratsiia*, September.

————. 1969. "Otnosit'sia k vodnym bogatstvam strany po Leninski" [Concern for the water riches of the country according to Leninist principles]. *Priroda*, December.

Bestuzhev-Lada, I. V. 1969. "Rekonstruktsiia planety: Proekty i Prognozy" [Reconstruction of the planet: projects and prognosis]. *Iunost'*. December.

Bobrov, S. N. 1961. "The Transformation of the Caspian Sea." *Soviet Geography: Review and Translation*, No. 7, p. 47.

Bogever, F. M., Prosenkov, V. L., and Iazvin, L. S. 1966. "Podzemnye vody Moskvy i podmoskov'ia" [Underground water of Moscow and the area around Moscow]. *Gorodskoe khoziaistvo Moskvy*, October.

Borisovich, G., and Vain, A. 1969. "Voprosy spetsializ-atsii v khimicheskoi industrii" [Problems of specialization in the chemical industry]. *Planovoe khoziaistvo*, February. I am grateful to Martin Spechler for this citation.

Braginskii, L. V. 1969. *Finansy melioratsii* [Financing land reclamation]. Moscow: Finansy.

Bryson, Reid A. 1971. "Atmospheric Imbalances and Pollution." Paper presented at International Conference on Environmental Future, Jyvaskyla, Finland, June 29.

Buchanan, J. M. 1969. "External Diseconomies, Corrective Taxes, and Market Structure." *American Economic Review*, March.

Buianovskaia, A. A. 1969. "O komissii po razrabotke problem okhrany prirodnykh vod AN SSSR" [About the Commission of the Academy of Sciences of the USSR for working out the problems of protecting inland fresh water]. *Priroda*, December.

Buiantuev, B. R. 1960. *K narodnokhoziaistvennym problemam Baikala* [In regard to the economic problems of Baikal]. Ulan-Ude: Buriatskoe knizhnoe izdatel'stvo.

Buiantuev, B. R., Galazii, G. I., Krotov, V. A., and Shotskii, V. P. 1962. "Problemy kompleksnogo ispol'zovaniia i okhrany prirodnykh resursov ozera Baikal" [Questions about the overall use and protection of the natural resources of Lake Baikal]. *Doklady instituta geografii Sibiri i dal'nego vostoka*, Irkutsk: Akademiia nauk SSSR, Sibirskoe otdelenie, No. 2.

Buinevich, D. V. 1969. "Perspektivy Kara-Bogaz-Gola" [Outlook for Kara-Bogaz-Gol]. *Priroda*, No. 5.

Bulavin, B. P. 1961. "Opolzni na Chernomorskom poberezh'e Kavkaza" [Landslides on the Black Sea coast of the Caucasus]. *Priroda*, June.

Campbell, Robert W. 1968. *The Economics of Soviet Oil and Gas*. Baltimore: The Johns Hopkins Press.

Cerkinskij, S. N. 1965. "The Basic Principles and Criteria for the Control of Water Pollution and the Possibilities of Using Them in Practice." Working Paper no. 6. Expert Committee on Water Pollution Control, Geneva, April 6–12, 1965, mimeographed.

Chebotarev, O. V. 1970. "Proizvodstvennoe zagriaznenie atmosfery" [Industrial pollution of the atmosphere]. *Zhurnal vsesoiuznogo khimicheskogo obshchestva im. D. I. Mendeleeva*, Vol. 15, No. 5.

Chernyshev, F. V. 1961. "Piatigor'e" [Piatigore mountains]. *Priroda*, December.

Chvanov, A. A. 1971. "Bol'shie zadachi bol'shogo goroda" [Major tasks in the large city]. *Vodosnabzhenie i sanitarnaia tekhnika*, February.

Commoner, Barry, Corr, Michael, and Stamler, Paul J. 1971. "The Causes of Pollution." *Environment*, Vol. 13, No. 3, April.

Council on Environmental Quality (CEQ). 1971. *The Second Annual Report on Environmental Quality*. Washington, D.C.: U.S. Government Printing Office.

Detwyler, Thomas R. 1971. *Man's Impact on Environment*. New York: McGraw-Hill Book Co.

D'iachkov, A. V. 1967. "Razvitie Moskovskogo vodoprovoda za 50 let" [The development of Moscow's water distribution system during the last 50 years]. *Vodosnabzhenie i sanitarnaia tekhnika*, November.

Dobrovol'skii, A. D., Kosarev, A. N., and Leont'ev, O. K., eds. 1969. *Kaspiiskoe more* [The Caspian Sea]. Moscow: Izdatel'stvo Moskovskogo universiteta.

Dolgullevich, M. I. 1966. "Pyl'nye buri na Ukraine" [Dust storms in the Ukraine]. *Izvestiia akademii nauk, seriia geografiia*, No. 1.

Dolgushin, I. Iu. 1969. "The Effect of Climatic Fluctuation on the Physical Environment and Conditions of Economic Development of the Middle Ob District." *Soviet Geography: Review and Translation*, No. 6, June.

Drachev, S. M., and Sinel'nikov, V. Ye. 1968. "Protection of Small Rivers Under Intensive Use, Taking the Moskva River as an Example." *Soviet Hydrology: Selected Papers*, No. 3.

Dzens-Litovskii, A. I. 1968. "Sokhranit' lechebnye pliazhi Kryma'" [To preserve the medicinal beaches of the Crimea]. *Priroda*, May.

Egorov, E. N. 1962. "Priostanovit' razrushenie iuzhnykh pliazhei" [Stop the destruction of the southern beaches]. *Priroda*, March.

"Ekonomicheskaia otsenka prirodnykh resursov" 1969. [An economic valuation of natural resources]. *Voprosy ekonomiki*, January.

Eliseev, N., and Kondratenko, A. 1968. "Nashe krovnoe delo" [Our sacred duty]. *Kommunist*, No. 5, March.

Engels, Frederick. 1940. *Dialectics of Nature*. New York: International Publishers.

———. 1955. *Dialektika prirody* [Dialectics of nature]. Moscow: Gospolitizdat.

Fedenko, N. 1966. "Moskva-reka dolzhna byt' chistoi" [The Moscow River must be clean]. *Priroda*, December.

Feitel'man, N. 1968. "Ob ekonomicheskoi otsenke mineral'nykh resursov" [About the economic valuation of mineral resources]. *Voprosy ekonomiki,* November. I am indebted to Robert Campbell for this citation.

Fonselius, Stig H. 1970. "Stagnant Sea." *Environment,* Vol. 12, No. 6, July–August.

Friedman, Milton. 1962. *Capitalism and Freedom.* Chicago: University of Chicago Press.

Gabyshev, K. E. 1969. "Ekonomicheskaia otsenka prirod-nykh resursov i rentnye platezhi" [The economic valuation of natural resources and rent payments]. *Vestnik Moskovskogo universiteta, seriia ekonomika,* No. 5, pp. 17–23.

Galanin, P. I. 1967. "Razvitie Moskovskoi kanalizatsii za 50 let" [The development of the Moscow sewer system during the last 50 years]. *Vodosnabzhenie i sanitarnaia tekhnika,* November, pp. 34–39.

Galazii, G. I. 1968. *Baikal i problema chistoi vody v Sibiri* [Baikal and Problems of Pure Water in Siberia]. Irkutsk: Akademiia nauk SSSR.

Galazii, G. I., and Novoselov, E. N. 1964. *Baikal.* Irkutsk: Limnological Institute of the Siberian Division of the Academy of Sciences.

Gal'tsov, A. Ia. 1964. "Razvitie vodosnabzheniia i ratsional'noe ispol'zovanie vody" [Development of water supply and the rational use of water]. *Gorodskoe khoziaistvo Moskvy,* February.

Gambarian, M. E. 1968. "Bioproduktivnost' ozera Sevan" [The bioproductivity of Lake Sevan]. *Priroda,* August.

Geller, S. Yu. [Iu.]. 1962. "On the Question of Regulating the Level of the Caspian Sea." *Soviet Geography*: *Review and Translation*, No 1.

―――. 1967. "Problema Aral'skogo moria" [The problem of the Aral Sea]. *Izvestiia akademii nauk, seriia geografiia*, No. 6.

Gerasimov, I. P. 1962a. "Soviet Geographic Science and Problems of the Transformation of Nature." *Soviet Geography*: *Review and Translation*, No. 1.

―――. 1962b. "Reducing the Dependence of Soviet Agriculture on Natural Elements to a Minimum." *Soviet Geography*: *Review and Translation*, No. 2.

―――. 1968 "Basic Problems of the Transformation of Nature." *Soviet Geography*: *Review and Translation*, No. 6.

―――. 1969. "Nuzhen general'nyi plan preobrazovaniia prirody nashei strany" [A general plan for the transformation of nature in our country is needed]. *Kommunist*, No. 2, January.

―――. 1970. "Futurology in Soviet Geography." *Soviet Geography*: *Review and Translation*, No. 7, September.

―――. 1971. "Scientific Technical Progress and Geography." *Soviet Geography*: *Review and Translation*, No. 4, April.

Goldman, Marshall I. 1967. *Controlling Pollution*: *The Economics of a Cleaner America*. Englewood Cliffs, N.J.: Prentice-Hall.

―――. 1972. *Ecology and Economics*: *Controlling Pollution in the 70's*. Englewood Cliffs, N.J.: Prentice-Hall.

Gordeev, Iu. I. 1963. "Opravdano li sozdanie Nizhne-

Obskogo moria" [Is the creation of the Lower Ob Sea justified?]. *Priroda*, June.

Gorin, G. S. 1968. "Iz istorii goroda. Moskovskii vodoprovod" [From the city's history. Moscow's water distribution system]. *Gorodskoe khoziaistvo Moskvy*, March.

Grin, A. M., and Koronkevich, N. I. 1963. "Principles of Construction of Long Term Water-Management Balance." *Soviet Geography: Review and Translation*, No. 3, March.

Gumilev, L. N. 1964. "Khazaria and the Caspian." *Soviet Geography: Review and Translation*, No. 6, June.

Gussak, L. A. 1964. "Vykhlopnye gazy mozhno obezvredit'" [Exhaust gases can be rendered harmless]. *Priroda*, March.

Gustafson, Thane. 1970. "Politics of Pollution in the Soviet Union." Unpublished paper, May 15, 1970.

Ianovskii, A. G. 1964. "O sisteme promyshlennogo vodosnabzheniia" [About the system of industrial water supply]. *Gorodskoe khoziaistvo Moskvy*, February.

Ianshin, A. L. 1965. "Moral'nyi dolg nashego pokoleniia" [The moral duty of our generation]. *Priroda*, July.

Iordanskiy, A. 1970. "The Caspian Calls for Help." *Khimiya i zhizn'*, No. 1, pp. 49–55. Reprinted in *Joint Press Review Service*, No. 120, August 12, 1970.

Ivanchenko, A. A., ed. 1969. *Ekonomicheskie problemy razmeshcheniia proizvoditel'nykh sil SSSR* [Economic problems of location of the productive forces in the USSR]. Moscow: Nauka.

Ivanov, I. T. 1970. "Vodosnabzhenie i kanalizatsiia gorodov SSSR" [Water supply and the sewer system in cities of the USSR]. *Vodosnabzhenie i sanitarnaia tekhnika*, April.

Ivanovskaia, T. M. 1970. "Sostoianie slizistoi obolochki dykhatel'nykh putei u rabochikh asbestovoi promyshlennosti" [The state of the mucous membrane in the respiratory tract of workers in the asbestos industry]. *Gigiena truda i professional'nye zabolevaniia*, No. 2.

Kadulin, V. 1967. "Primor'e-krai bogateishikh vozmozhnostei" [The Primore region—a very rich opportunity]. *Kommunist*, No. 11, July, pp. 81–91.

Kaliuzhnyi, D. N., Kostovetskii, Ia. I., and Ianysheva, N. Ia. 1968. *Sanitarnaia okhrana atmosfernogo vozdukha i vodoemov ot vybrosov i otkhodov predpriiatii chernoi metallurgii* [The sanitary protection of the atmospheric air and water courses from emissions and wastes of ferrous metal enterprises]. Moscow: Meditsina.

Kaposhin, T. S. 1971. "Razvitie vodoprovodno-kanalizatsionnogo khoziaistva v 1966–1970 gg" [The development of water distribution sewage systems in 1966–1970]. *Vodosnabzhenie i sanitarnaia tekhnika*, February.

Kapp, K. William. 1963. *Social Costs of Business Enterprise*. New York: Asia Publishing House.

Kazantsev, N. D. 1967. *Pravovaia okhrana prirody v SSSR* [The legal protection of nature in the USSR]. Moscow: Znanie, series 17.

Kes', A. S. 1967. "Problemy preobrazovaniia prirody srednei Azii" [Problems of the transformation of nature in Central Asia]. *Izvestiia akademii nauk, seriia geografiia*, No. 6.

Khachaturov, T. 1969. "Ob ekonomicheskoi otsenke prirodnykh resursov" [About the economic valuation of natural resources]. *Voprosy ekonomiki,* January, pp. 66–74.

Khodyrev, N. A. 1964. "Formy razrusheniia morskikh beregov v Gruzii" [The forms of destruction of the sea coast in Georgia]. *Priroda,* July.

Kneese, Allen V. 1962. *Water Pollution,* Washington, D.C.: Resources for the Future.

―――. 1964. *The Economics of Regional Water Quality Management.* Baltimore: The Johns Hopkins Press.

Kneese, Allen V., and Smith, Stephen C., eds. 1966. *Water Research.* Baltimore: The Johns Hopkins Press.

Kolbasov, O. S. 1965. *Legislation on Water Use in the U.S.S.R.,* translated by Rosemary J. Fox. Moscow: Iurizdat.

Kop'ev, S. F. 1967. "Razvitie teplosnabzheniia v SSSR" [The development of the central heating supply system in the USSR]. *Vodosnabzhenie i sanitarnaia tekhnika,* November.

Korzhenevskii, I. B., Loenko, A. A., and Cherevkov, V. A. 1961. "Sud'ba pliazhei iuzhnogo berega Kryma" [The fate of the beaches of the southern Crimean coast]. *Priroda,* February.

Kosarev, Alexei. 1970. "The Diminishing Southern Sea." *Soviet Life,* July.

Kozlov, V. M., Zykova, A. S., Zhakov, Iu. A., and Iambrovskii, Ia. M. 1970. "Radiatsionnaia bezopasnost' naseleniia v raione raspolozheniia atomnykh elektrostantsii" [Radioactive danger to the population in the

region where atomic electric power stations are located].
Gigiena i sanitariia, No. 4.

Kravinko, I. V. 1961. "Chernye buri" [Black dust
storm]. *Priroda*, December.

Kriazhev, V. G. 1966. *Vnerabochee vremia i sfera obsluzhi-
vaniia* [Free time and the service sector]. Moscow:
Ekonomika.

Kuznetsov, I. A. 1970. *Izuchenie i ispol'zovanie vodnykh
resursov SSSR* [Study and use of water resources of the
USSR]. Moscow: Nauka.

Kuznetsov, I. A., and Tikhomirov, F. K. 1965. "Ladoz-
hsko-Kaspiiskii vodnyi trakt" [The Ladozga-Caspian
water route]. *Priroda*, January.

Lamakin, V. V. 1965. *Po beregam i ostrovam Baikala*
[Along the shores and islands of Baikal]. Moscow:
Nauka.

Lange, Oscar, and Taylor, Fred M. 1938. *On the Economic Theory of Socialism*. Minneapolis: University of
Minnesota Press.

Leporskii, D. V., and Nazarov, I. A. 1969. "Oroshenie
stochnym i vodami na Ukraine" [Irrigation with sewage
water in the Ukraine]. *Gidrotekhnika i melioratsiia*, April.

Lesnikova, N. P. 1970. "Pravovye problemy okhrany
prirody v SSSR" [Legal questions of the protection of
nature in the USSR]. *Priroda*, January.

Lesnikova, N. P., and Fedenko, N. F. 1968. "Vsemernaia
i polnaia zashchita prirody Kryma" [The comprehensive
and complete defense of nature in the Crimea]. *Priroda*,
December.

Livchak, I. F., and Kop'ev, S. F. 1969. "Puti dal'nei-shego sovershenstvovaniia tekhniki otopleniia i teplos-nabzheniia" [The road to further improvement of the technique of heating and central supply of steam heat]. *Vodosnabzhenie i sanitarnaia tekhnika*, September.

Loiter, M. 1967. "Ekonomicheskie mery po ratsional'no-mu ispol'zovaniiu vodnykh resursov" [Economic measures for the rational use of water resources]. *Voprosy ekonomiki*, December, pp. 75–86.

"Luchshe ispol'zovat' rezervy vodosnabzheniia goroda." 1964. [A better use of the reserves of the city water supply]. *Gorodskoe khoziaistvo Moskvy*, February.

L'vovich, A. I. 1963. "Problema okhrany rek i vodoemov ot zagriazneniia stochnymi vodami" [Problems of preserving rivers and water courses from pollution of water sewage]. *Izvestiia akademii nauk, seriia geografiia*, No. 3.

L'vovich, M. I. 1962. "Complex Utilization and Protection of Water Resources." *Soviet Geography: Review and Translation*, No. 10, December.

————. 1963a. *Chelovek i vody*, [Man and water]. Moscow: Geografgiz.

————. 1969a. "Scientific Principles of the Complex Utilization and Conservation of Water Resources." *Soviet Geography: Review and Translation*, No. 3, March.

————, ed. 1969b. *Vodnyi balans SSSR i ego preobrazovanie.* [The water balance of the USSR and its transformation]. Moscow: Nauka.

L'vovich, M. I., Grin, A. M., and Dreier, N. N. 1963. *Osnovy metoda izucheniia vodnogo balansa i ego preobrazovaniia* [Principles of the method of studying the water balance and its transformation]. Moscow: Akademiia nauka, SSSR.

Lysenko, V. A. 1969. "Okhrana prirody—nashe krovnoe delo" [The protection of nature is our sacred duty]. *Kommunist Ukrainy*, No. 1, January.

Makeenko, M. 1968. "Preodolenie sotsial'no-ekonomich-eskikh razlichii mezhdu gorodom i derevnei" [Overcoming the social-economic differences between the city and the countryside]. *Voprosy ekonomiki*, August.

Malinkevich, L. B., and Belianov, B. A. 1966. "Kaspii glubzhe kilometra" [The Caspian is more than a kilometer in depth]. *Priroda*, August.

Markizov, V. I. 1964. "Vazhnaia otrasl' gorodskogo khoziaistva" [An important branch of the city's economy]. *Gorodskoe khoziaistvo Moskvy*, February.

————. 1966. "Kachestvo raboty i ekonomika vodoprovoda i kanalizatsii" [The quality of work and the economics of water distribution and sewer systems]. *Gorodskoe khoziaistvo Moskvy*, October.

Marx, Karl. 1959a. *Capital*. Moscow: Foreign Languages Publishing House, Vol. 1.

————. 1959b. *Capital*. Moscow: Foreign Languages Publishing House, Vol. 3.

Maryan, Yuri, 1968. "The Lake Sevan Problem." *Soviet Life*, March.

"Masshtaby nashego rosta" [The scale of our growth]. *Vodosnabzhenie i sanitarnaia tekhnika*, November.

Mateev, A. V. 1969. "Otnoshenie V.I. Lenina k voprosam uluchsheniia inzhenernogo blagoustroistva naselennykh mest" [The attitude of V. I. Lenin toward questions of the improvement of population centers through engineering]. *Vodosnabzhenie i sanitarnaia tekhnika*, May, pp. 1–4.

Mazanova, M. 1969. "Territorial'nye proportsii razvitiia ekonomiki" [Territorial proportions in the development of the economy]. *Planovoe khoziaistvo*, February.

Medunin, A. E. 1969. "Vliianie nauchno-tekhnicheskoi revoliutsii na prirod zemli" [The influence of the scientific-technological revolution on the earth's nature]. *Voprosy filosofii*, March.

Mezentsev, V. S. 1964. "The Natural Moisture Balance of the West Siberian Plain and the Lower Ob Problem." *Soviet Geography: Review and Translation*, No. 5, May.

Micklin, Philip P. 1967. "The Baykal Controversy: A Resource Use in Conflict in the USSR." *Natural Resources Journal*, No. 4.

————. 1969. "Soviet Plans to Reverse the Flow of Rivers: The Kama-Vychegda-Pechora Project." *The Canadian Geographer*, No. 3.

Ministerstvo Vneshnei Torgovli SSSR. 1970. *Vneshniaia torgovlia SSSR za 1969 god* [The foreign trade of the USSR for 1969]. Moscow: IMO.

Mishan, E. J. 1967. *The Costs of Economic Growth*. New York: Praeger.

————. 1969. *Technology and Growth: The Price We Pay*. New York: Praeger.

Nagibina, T. 1961. "Organization of Water Pollution Control Measures in the USSR and Eastern European Countries." Conference on Water Pollution Problems in Europe. Geneva: United Nations Printing Office, Vols. 1–3.

Nekrasov, N. 1966. "Nauchnye problemy razrabotki general'noi skhemy razmeshchenia proizvoditel'nykh sil SSSR" [Scientific problems in developing a general scheme for the location of the productive forces of the USSR]. *Voprosy ekonomiki*, September, pp. 3–14.

Nuttonson, M. Y., ed. 1969a. *Atmospheric and Meteorological Aspects of Air Pollution. AICE Survey of USSR Air Pollution Literature*. Silver Spring, Md.: American Institute of Crop Ecology, Vol. 1.

————, ed. 1969b. *Effects and Symptoms of Air Pollutes on Vegetation: Resistance and Susceptibility of Different Plant Species in Various Habitats in Relation to Plant Utilization for Shelter Belts and as Biological Indicators. AICE Survey of USSR Air Pollution Literature*. Silver Spring, Md.; American Institute of Crop Ecology, Vol. 2.

————, ed. 1970a. *The Susceptibility or Resistance to Gas and Smoke of Various Arboreal Species Grown Under Diverse Environmental Conditions in a Number of Industrial Regions of the Soviet Union. AICE Survey of USSR Air Pollution Literature*. Silver Spring, Md.: American Institute of Crop Ecology, Vol. 3.

————, ed. 1970b. *Meteorological and Chemical Aspects of Air Pollution: Propagation and Dispersal of Air Pollutants in a Number of Areas in the Soviet Union. AICE Survey of USSR Air Pollution Literature*. Silver Spring, Md.: American Institute of Crop Ecology, Vol. 4.

Osipova, G. A., Ozherel'eva, M. N., and Usacheva, A. A. 1969. "Sostoianie proektirovaniia, stroitel'stva i ekspluatatsii kanalizatsionnykh ochistnykh sooruzhenii nebol'shoi proizvoditel'nosti" [The state of planning, construction and operation of the sewer purification equipment of limited productivity]. *Vodosnabzhenie i sanitarnaia tekhnika*, September.

"Otsenka prirodnykh resursov." 1968. [Valuation of natural resources]. *Voprosy geografii*, No. 78. Moscow: Mysl'.

Ovsiannikov, N. G. 1964. "Berech' i umnozhat' vodnye bogatstva strany" [Guard and augment the water riches of the country]. *Priroda*, March.

Oziranskii, S. 1968. "Plata za vodnye resursy" [Payment for water resources]. *Planovoe khoziaistvo*, September.

Panov, D. G., and Mamykina, V. A. 1961. "Mozhno li priostanovit' razrushenie beregov Azovskogo moria?" [Is it possible to stop the destruction of the shores of the Azov Sea?] *Priroda*, May.

Petrianov, I. 1969. "Vozdushnaia sreda: problemy i perspektivy ee zashchity" [Problems of the atmosphere and the prospects for its protection]. *Kommunist*, No. 11, July, pp. 71–80.

Pigou, A. C. 1912. *Wealth and Welfare*. London: Macmillan and Co., Ltd.

———. 1928. *A Study in Public Finance*. London: Macmillan and Co., Ltd.

Powell, David E. 1971. "The Social Costs of Modernization: Ecological Problems in the USSR." *World Politics*, July.

Pridatchenkov, V., and Keylin, D. 1971. "The Air Must Be Clean." *Promyshlennost' Belorussii*, No. 3, pp. 63–64. Translated in *Joint Publication Research Service, Resources*, No. 183, May 4, 1971.

Proshchenko, V. F. 1965. "Pyl'naia buria zimoi" [Winter dust storms]. *Priroda*, February.

Romanenko, A. M. 1967. "Plata za potreblenie i sbros vody" [Payment for the use and discharge of water]. *Gidrotekhnika i melioratsiia,* June.

Rossolimo, L. 1966. *Baikal.* Moscow: Nauka.

Samuelson, Paul A. 1970. *Economics,* 8th edition. New York: McGraw-Hill Book Co.

Sarukhanov, G. L. 1961. "Pechora-Kaspii" [Pechora-Caspian]. *Priroda,* July.

Sennikov, V. A. 1968. "Baikal." *Geografiia v shkole,* No. 5.

Serova, O., and Sarkisian, S. 1961. *Zhemchuzhina vostochnoi Sibiri* [The pearl of Eastern Siberia]. Ulan-Ude: Buriatskoe knizhnoe izdatel'stvo.

Shabad, Theodore. 1951. *Geography of the USSR.* New York: Columbia University Press.

————. 1969. *Basic Industrial Resources of the U.S.S.R.* New York: Columbia University Press.

Shabalin, A. F. 1970. "Leninskii plan industrializatsii strany i razvitie promyshlennogo vodosnabzheniia" [Lenin's plan for the industrialization of the country and the development of industrial water supply]. *Vodosnabzhenie i sanitarnaia tekhnika,* April.

Shalamberidze, O. P., and Pirtskhalava, S. E. 1970. "K voprosu ozdorovleniia atmosfernogo vozdukha v Gruzinskoi SSSR" [Toward the question of purifying the atmosphere in the Georgian Republic]. *Gigiena i sanitariia,* No. 4.

Shishkin, N. I. 1962. "On the Diversion of the Vychegda and Pechora Rivers to the Basin of the Volga." *Soviet Geography: Review and Translation,* No. 5.

Shitunov, F. Ia. 1969. "Budut li zhit' gory?" [Will the mountains survive?] *Priroda*, September.

Shkatov, V. 1969. "Prices on Natural Resources and the Problem of Improving Planned Price Formation." *Problems of Economics*, No. 2, June.

Shnitnikov, A. V. 1968. "Veroiatnye tendentsii kolebanii vodnosti territorii SSSR" [Probable trends in the variations of water resources in the USSR]. *Voprosy Geografii*, No. 73.

Shul'ts, V. L. 1968. "The Aral Sea Problem." *Soviet Hydrology: Selected Papers*, No. 5.

Simonov, Anatoli. 1970. "Cleaning Up the Sea." *Soviet Life*, January.

Sokolovskii, M. S., Gabinova, Zh. L., Popov, B. V., and Kachor, L. F. 1965. *Sanitarnaia okhrana atmosfernogo vozdukha Moskvy* [Sanitary protection of Moscow's atmosphere]. Moscow: Meditsina. I am grateful to Mark Field for drawing this source to my attention.

Soloukhin, Vladimir. 1967. *A Walk in Rural Russia*. New York: E. P. Dutton.

Stas' [Stass], I. I. 1968. "Kompleksnoe osvoenie Kara-Bogaz-Gola" [The overall development of Kara-Bogaz-Gol]. *Priroda*, September.

———. 1970. "Saving the Caspian from Drying Up." *Sputnik*, May.

Streeten, Paul. 1971. "Cost Benefit and Other Problems of Method." Paper presented at the conference of Political Economy of Environment: Problems of Method, Paris, July 5–9, 1971.

Sukhotin, Iu. 1967. "Ob otsenkakh prirodnykh resursov" [About evaluating natural resources]. *Voprosy ekonomiki*, December, pp. 87–98.

————. 1968. "Concerning Evaluation of Natural Resources." *Problems of Economics*, July.

Sushkov, T. S. 1969. "Pravovaia okhrana prirody" [The
legal protection of nature]. *Sovetskoe gosudarstvo i pravo*,
May.

Timoshchuk, V. I. 1968. "Prirodnyi strontsii v Kaspii-
skom more" [Natural strontium in the Caspian Sea].
Priroda, January.

Tolstoi, M. P., and Bondarev, V. P. 1964. "Tsennoe
syr'e dlia mineral'nykh udobrenii. Shiroko ispol'zovat'
promyshlennye otkhody" [Industrial waste widely used
as a valuable source for mineral fertilizer]. *Priroda*,
August.

Trofiuk, A. A., and Gerasimov, I. P. 1965. "Sokhranit'
chistotu vod ozera Baikala!" [Preserve the pure water of
Lake Baikal!]. *Priroda*, November.

Troshkina, E. S., and Volodicheva, N. A. 1968. "Otsenka
lavinnoi opasnosti gornykh raionov pribaikal'ia" [Esti-
mating the danger of landslides in the mountainous
region around Lake Baikal]. *Vestnik Moskovskogo universiteta, seriia geografiia*, No. 5.

Tsentral'noe statisticheskoe upravlenie. 1968. *Narodnoe
khoziaistvo SSSR v 1967 gody* [The National Economy
of the USSR in 1967]. Moscow: Statistika.

Tsuru, Shigeto, ed. 1970. *Proceedings of International Sym-
posium on Environmental Disruption*. Tokyo: Asahi Evening
News.

————. 1971. "In Place of GNP." Unpublished article, February.

U.S. Bureau of the Census. 1970. *Statistical Abstract of the United States:* 1970, 91st edition. Washington, D.C.: United States Government Printing Office.

Vasil'eva, E. M., Tamm, O. M., and Ianes, Kh. Ia. 1970. "Zagriaznenie atmosfernogo vozdukha Tallina vykhlopnymi gazami avtotransporta" [Air pollution in Tallin from the exhaust gases of motor vehicles]. *Gigiena i sanitariia,* No. 2.

Vendrov, S. L. 1964. "Water Management Problems of Western Siberia." *Soviet Geography: Review and Translation,* No. 5, May.

————. 1965. "A Forecast of Changes in Natural Conditions in the Northern Ob' Basin in Case of Construction of the Lower Ob' Hydro Project." *Soviet Geography: Review and Translation,* No. 10, December.

————. 1966. "Dinamika beregov krupnykh vodoemov v sviazi s ispol'zovaniem vodnykh resursov" [Coastal dynamics of big reservoirs in connection with the use of water resources]. *Izvestiia akademii nauk, seriia geografiia,* No. 2.

Vendrov, S. L., Gangardt, G. G., Geller, S. Yu., Korenistov, L. V., and Sarukhanov, G. L. 1964. "The Problem of Transformation and Utilization of the Water Resources of the Volga River and the Caspian Sea." *Soviet Geography: Review and Translation,* No. 7, September.

Vendrov, S. L., and Kalinin, G. P. 1960. "Surface Water Resources of the USSR: Their Utilization and Study." *Soviet Geography: Review and Translation,* June.

Vitt, M. 1970. "Ob ekonomicheskikh stimulakh ratsional'nogo ispol'zovaniia prirodnykh resursov" [About the economic stimulus for the rational use of natural resources]. *Planovoe khoziaistvo*, July.

Weinberger, Leon W., Stephan, David G., and Middleton, Francis M. 1966. "Solving our Water Problems—Water Renovation and Reuse." *Annals of the New York Academy of Sciences*, Vol. 136, Art. 5, July 8.

Zenkovich, V. P. 1967. "Comments." *Vestnik, akademiia nauk, SSSR*, No. 7.

Zenkovich, V. P., and Zhdanov, A. M. 1960. "Pochemu ischezaiut Chernomorskie pliazhi" [Why the Black Sea beach is disappearing]. *Priroda*, October.

Zhakov, Z. O. 1964. "The Long-Term Transformation of Nature and Changes in the Atmosphere Moisture Supply of the European Part of the U.S.S.R." *Soviet Geography: Review and Translation*, No. 3, March.

Zhirkov, K. F. 1964. "Dust Storms in the Steppes of Western Siberia and Kazakstan." *Soviet Geography: Review and Translation*, May.

Zhukov, A. I. 1970. "Ochistka stochnykh vod i zashchita vodoemov ot zagriazneniia" [The treatment of sewage and the protection of water courses from pollution]. *Vodosnabzhenie i sanitarnaia tekhnika*, April.

Zhukov, A. I., and Mongait, I. L. 1967. "Dostizheniia v oblasti ochistki promyshlennykh stochnykh vod" [Achievements in the area of treatment of industrial sewage]. *Vodosnabzhenie i sanitarnaia tekhnika*, November, pp. 13–17.

Zhuvkovich, L. A. 1967. "Skorost' zapolneniia Kamsko-Pechorsko-Vychegodskogo vodokhranilishcha" [The speed of filling the Kama-Pechora-Vychegda water reservoir]. *Vestnik Moskovskogo universiteta, seriia geografiia*, No. 2.

Zile, Zigurds L. 1970a. "Lenin's Contribution to Law: The Case of Protection and Preservation of the Natural Environment." Prepared for Lenin and Leninism: A Centenary Conference on State, Law and Society, at Oklahoma State University, Stillwater, Okla., April 2–22, 1970, mimeographed.

————. 1970b "The Social Costs of Socialist Development: Environmental Deterioration in the Soviet Union." Prepared for the Ninth Annual Bi-State Slavic Conference (Kansas-Missouri) at the University of Kansas, Lawrence, Kan., November 6–7, 1970, mimeographed.

Zinger, N. M. 1970. "Teplosnabzhenie abonentov v avariinykh usloviiakh" [Steam heat supply to customers in emergency situations]. *Vodosnabzhenie i sanitarnaia tekhnika*, October.

Zykova, A. S., Telushkina, E. L., Rublevskii, V. P., Efremova, G. P., and Kuznetsova, G. A. 1970. "Soderzhanie iskusstvennykh radioaktivnykh izotopov v atmosfernom vozdukhe Moskvy v 1962–1967 gg" [The content of artificial radioactive isotopes in Moscow's air in 1962–1967]. *Gigiena i sanitariia*, No. 4.

Newspapers and Journals
Bakinskii rabochii (Bak Rab)

Boston Globe (BG)

Business Week

Current Digest of the Soviet Press (CDSP)

Ekonomicheskaia gazeta (EG)

Federal Register

Gidrotekhnika i melioratsiia

Izvestiia (Iz)

Kazakhstanskaia pravda (Kaz Prav)

Kommunist (Kom)

Kommunist tadzhikistana (Kom Tad)

Komsomolskaia pravda (Kom Prav)

Krasnaia zvezda (Kras zvez)

Krokodil

Literaturnaia gazeta (LG)

Meditsinskaia gazeta (Med Gaz)

New York Times (NYT)

Pravda (Prav)

Pravda ukrainy (Prav Uk)

Pravda vostoka (Prav Vos)

Priroda

Rabochaia gazeta (Rab Gaz)

Sea Secrets, No. 6, 1970

Selskaia zhizn' (Sel Zh)

Sotsialisticheskaia industriia (Sot In)

Sovetskaia belorussiia (Sov Bel)

Sovetskaia estoniia (Sov Est)

Sovetskaia kirgiziia (Sov Kir)

Sovetskaia latviia (Sov Lat)

Sovetskaia litva (Sov Lit)

Sovetskaia moldaviia (Sov Mold)

Sovetskaia Rossiia (Sov Ros)

Soviet Geography: *Review and Translation (SGRT)*

Soviet Life (SL)

Soviet News (SN)

Trud

Turkmenskaia iskra (Turk Isk)

Vodosnabzhenie i sanitarnaia tekhnika (VST)

Voprosy ekonomiki (VE)

Wall Street Journal (WSJ)

Zaria vostoka (Zar Vos)

Zhurnal vsesoiuznogo khimicheskogo obshchestva im. D. I. Mendeleeva, Vol. XV, No. 5, 1970.

Index

Abkhazia, 159, 168
Abramov, B. S., 81, 84
Academy of Science, 159, 183, 241
Accountability, 188–189
Adler, 155, 158
Aeration, 70, 99, 200
Aerosols, 128, 143
Agriculture, 6, 39, 56, 60, 63, 81,
    111, 113, 114, 115, 117, 118,
    167, 171–173, 218, 228, 257, 258,
    265–266, 273, 282, 283, 302–303,
    316, 320. *See also* Kolkhozy;
    Peasant; Sovkhozy
Agronomist, 63
Air, 5–7, 20, 23, 28, 35, 37–38, 44,
    46, 62, 103–104, 111, 116, 121–
    150, 152, 165, 173–174, 180–181,
    187, 189, 201, 276–277, 287. *See
    also* Atmosphere
  circulation of, 142
  concentration limits in, 24, 25
    (Table 1.1)
  free good, 47–48
  inversion, 141–142
  laws, 295, 297–299, 302
  legislation, 27, 29–31, 38
Akimovich, N. N., 138
Alienation of man, 11

Alkaline, 200–201
Allen, Gary, 3
Allocation of resources, 279
Alma-Ata, 60, 62, 118, 137, 145
Aluminum, 111
Amortization rate, 93
Amu-Bukhara Canal, 223, 224
    (map), 257
Amu Darya, 55, 56, 218, 223, 224
    (map), 235, 241, 248 (map), 250,
    257, 267
Angara River, 179, 184–185, 247
Apartments. *See* Housing
Apollov, B. A., 216–217, 226,
    232, 236, 237, 241, 251, 253,
    255
Aquaphilia, 221
Aral Sea, 7, 55–56, 80, 215–218,
    223, 224 (map), 225–226, 229,
    235–238, 240–242, 248 (map),
    249–250, 257, 265
Aral'sk, 236
Arctic Ocean, 79, 179, 246, 262,
    263, 269, 270, 286
Armand, D., 28–29, 35–37, 80, 110,
    140, 148, 168–171, 225, 256, 258,
    259–260, 274–275, 278, 286
Armenia, 65, 296

Artesian well, 93–96, 294, 303
Article 223 of Criminal Code of
    RSFSR, 37
Asbest, 142
Asbestos, 142
Ashkhabad, 224 (map), 235, 248
    (map)
Astara, 236
Astrakhan, 222 (map), 227 (map),
    235, 275, 294
Astrakhantsev, G., 115, 117–118
Astrakhantsev, V. I., 180, 182, 198,
    201–202
Aswan Dam, 234–235
Atlantic Ocean, 270
Atmosphere, 302, 307–308
Atomic Energy Commission, 192
Atomic power, 143–144, 289, 290
Atomic Test Ban Treaty, 128, 287
Atomic testing, 237
Automobiles, 37, 125, 130–134,
    136–137, 160, 264, 280–281
Azerbaijan, 38, 149, 296
Azov, Sea of, 79, 157, 221, 222
    (map), 245, 246, 248 (map), 250,
    286

Baikal. See Lake Baikal
Baikalsk Cellulose Carton fac-
    tory, 184, 189–193, 195–202,
    206–208
Baku, 93, 230
Balandina, V. A., 142
Balkhash (town), 59
Baltic Republic, 284
Baltic Sea, 79, 221, 251, 252
    (map), 284–285
Bannikov, A. G., 273–274
Barents Sea, 222 (map), 286
Barguzin National Preserve, 190,
    274
Batum, 156
Bazenkov, F. A., 256
Beach Laboratory, 159
Beaches, 72–73, 154–163, 298,
    300
Belianov, B. A., 216

Belorussia, 105, 297
Bel'tsakh, 105
Belyi Amur (fish), 235
Bendera, 106
Bering Strait, 269–270
Bestuzhev-Lada, I. V., 256–257,
    270
Black Sea, 32, 66, 72, 79, 154–163,
    165, 168, 221, 245–246, 248
    (map), 250, 270, 285, 298, 300
    salt content of, 246
Blasting, 229
Bobrov, S. N., 217–218, 228, 232,
    236, 241, 244–245
Bogatir' factory, 17
Bogdanov, B., 118
Bogever, F. M., 94
Bondarev, V. P., 283
Bonus, 65, 116–117
Borgustansk mountain range,
    164–165
Borisovich, G., 52
Bratsk, 185, 187–188, 207
Brest, 105
Brigade, 85–86
Bryson, Reid, 170
Buchanan, J. M., 21
Buianovskaia, A. A., 83
Buiantuev, B. R., 182–183, 190,
    195–196, 207
Buinevich, D. V., 228
Bukh Tarm, 203
Bulavin, B. P., 170
Bureaucracy, 275
"Bureaucratic confusion," 165
Buriat Republic, 182

Campaign, 147, 186
Campbell, Robert W., 133
Canal, 93, 106, 119, 221, 222
    (map), 223 n, 224 (map), 225,
    235, 247, 248 (map), 249–250,
    252 (map), 253, 254 (map), 256–
    257, 260–261, 265, 312, 315, 321–
    322
Cape Pitsunda, 155
Capital, 20, 92, 101, 103, 118, 146,

Capital (*continued*)
188, 281

Capitalism, 3, 4, 10, 23, 58, 71

Carbon monoxide, 131, 134

Caspian Sea, 7, 80, 215–219, 222
(map), 223, 224 (map), 225–
226, 227 (map), 228–235, 240–
246, 248 (map), 249–251, 252
(map), 269, 285–286
  legislation about, 229–230, 299
  strontium concentration level of,
  236–238

Caucasus, 163

Caviar, 233, 244
  synthetic, 244

Cellulose plant, 183–186, 189–190,
193–195, 201. *See also* Selenga
and Baikalsk

Central Asia, 79–80, 106, 218,
224 (map), 225, 249, 259, 262,
266

Central Committee of the Soviet
Communist Party, 208–209

Central heat and hot water, 275–
278. *See also* Teplovaia
Elektrotsentral plants

Central Ore Concentration Mill,
141

Checks and balances, 33–34, 73

Cheliabinsk, 137, 139, 145

Chelysukintsev Park, 141

Chemical industry, 6–7, 20, 62, 70,
101, 126, 129, 135, 137–138, 146–
147, 207, 278, 285

Chernyshev, F. V., 167

Cherporets Steel Mill, 111

Chief Administration for Sanitary
Epidemiological Supervision, 29,
122–123, 295

Chief Sanitary Inspector, 197,
298–299

Chistiakov, Nikolai, 186–187, 198,
201

Chlorination, 101

Cholera, 235

Chvanov, A. A., 91, 92, 102

Civil war, 88

Clean Air Act of 1970, 27

Climate, 56, 155, 180, 205, 218,
235, 240, 262–263, 305

Coal, 61, 127, 139

Codification, 38

Cold war, 270

Collective farm. *See* Kolkhozy

Collectivization, 170–171

Commoner, Barry, 83, 110–111

Communist Party. *See* Party

Compensation, 19

Concentration, 308

Concentration of water, 26

Conflict of interest, 36, 189–190

Conservation, 2, 11–12, 14, 16–19,
23, 28, 30, 33, 52, 54, 57, 62–63,
65–69, 72, 88, 101, 112–113, 115,
117, 153, 165, 172–173, 181, 183,
186, 188–189, 193, 195, 205, 245,
262, 264, 267, 273, 280–281, 287
  Law, 301–310, 312–317, 319–329

Conservationists. *See* Conservation

Consumer goods, 279–282

Conversion table, 331

Cost benefit analysis, 53–58, 172–
173, 175, 257, 266, 270

Costs, 19, 21, 45, 47–52, 54, 56–
58, 62, 65–66, 70, 74, 93–94, 103,
113, 115, 117–119, 132, 140,
149–150, 169, 172–173, 181, 188,
195, 199–200, 202, 204, 225, 232,
241, 255, 260, 280. *See also* Opportunity cost; Externalities

Council of Ministers, 39, 63, 114,
122, 208, 229

Crimes, 79, 155, 160–161

Criminal code, 34–35, 37–38

Criminal liability, 310

Criminal offense, 149

Dam, 57–62, 89–92, 102, 118, 156–
158, 160, 172, 220–221, 222
(map), 223, 224 (map), 233–234,
243–247, 248 (map), 253, 254
(map), 256, 258–261, 264–265,
269–270. *See also* Reservoir

Danube River, 270
DDT, 45
De-emphasis of consumer goods, 279–281
Defoliation, 289–290
Demchenko, Ya., 247
Department of Public Works mentality, 57–64
Depletion allowance, 174
Detergent, 102
Detwyler, Thomas R., 255
Developed countries, 5, 154, 268, 274, 289
Developing countries, 64, 75, 214, 220, 268–269, 280
Development, 15, 28, 88, 171, 182–185, 213, 217, 290, 301–302, 306, 327–328. *See also* Growth; Industrialization
D'iachkov, A. V., 84–85, 92, 94
Diesel fuel, 133
Discount rate, 53
Dnepr, 109, 265
Dneprovsk Basin, 80
Dnester River, 109, 160
Dobrovol'skii, A. D., 216–219, 232, 234
Dolgullevich, M. I., 171
Dolgushin, I. Iu., 259
Donbass Region, 69, 95, 105, 114, 265
Donets Basin, 79–80, 105, 127
Don River, 221, 222 (map), 248 (map), 250
Drachev, S. M., 88, 96, 102
Drainage, 31, 118, 235, 259, 294, 315, 320, 323, 327, 330
Drought, 29
Dubossar, 106
Dust, 126, 138, 141, 150, 165, 170–171, 203, 212, 238, 240
Dzens-Litovskii, A. I., 160
Dzhinal'sk mountain range, 164, 166

Economic factors, 4, 5, 44, 75
Economic free good, 40, 47–50,

Economic free good (*continued*) 72, 162–163, 174, 258, 260
Effluent, 3, 6, 19–20, 24, 50, 54, 83, 94, 97, 101–102, 107, 115–117, 137, 168, 182, 185, 190, 192, 196, 198–202, 228, 230–231, 276–277, 279, 282–283, 285, 304, 318, 320–321, 327
Egorov, E. N., 157, 159
Eighth Five-Year Plan, 92
Electricity, 6, 20, 39, 44, 60–62, 65–66, 73, 82, 95, 107, 126, 130, 136, 139, 142–144, 156–157, 185, 244, 259, 264, 275–276, 282. *See also* Hydroelectric
Electrostatic precipitators, 70
Eliseev, N., 170
Enforcement of pollution laws and controls, 27–29, 31–34, 36–38, 149, 161, 190, 231, 309–310, 313–314, 319
Engels, Frederick, 11–14, 18, 22, 213–214, 267
Entrepreneur, 22, 146–147
Erevan, 93
Erosion, 31, 66, 118, 157–158, 160, 162, 167–173, 190, 203, 209, 299, 302–303, 307, 315, 327–328, 330
Estonia, 30, 38, 105, 131, 149, 296
Eutrophication, 66
Evaporation, 111, 215, 218, 223, 228, 237, 241, 260–262, 320
Exports, 167–168
Externality, 19–22, 44–53, 71–74, 103, 140, 173, 264, 268, 281, 290
Extraction, 48, 50, 52

Factory close down, 36–37, 126, 278–279
Fedenko, N., 97
Federalism, absence in USSR, 33–34
Federenko, N. P., 114
Fedorova, Iu., 62–63
Feitel'man, N., 49
Fergana Valley, 223

Fertilizer, 244, 282, 302
Fiat automobile plant, 130
Filevskaia, 99
Filter, 69, 145, 148
Fines, 34–40
Fish, 28, 31, 34, 37, 56, 58–59, 69,
    81, 101, 109, 117, 168, 181–182,
    191, 203–204, 228, 232–235, 240,
    241, 243, 244, 245, 246, 269,
    285–286
  breeding, 244, 264
  breeding grounds, 234–235, 245,
    304
  catch, 232 (Table 7.3)
  fish gate, 233, 244
  laws for, 294–295, 299, 303–304,
    307, 309, 315–317, 320, 322–326
Five-year plans, 28, 64, 92, 123,
    169, 187, 255–256
Flooding, 57, 60–61, 73, 172, 225,
    234, 258, 260, 261, 262, 268, 303,
    315–316, 328. See also Reservoir
Fluoride, 141
Fluorine, 142
Fonselius, Stig H., 284
Food chain, 233
Foreign exchange, 167, 233. See
    also Caviar; Exports
Forest Laboratory of Gosplan, 124
Forests, 14, 17–18, 29–31, 62–63,
    123–124, 140–141, 152–155, 159,
    166–171, 181, 183–184, 187, 190,
    195, 197, 202–203, 205, 258, 275
  legislation, 293, 294–296, 299,
    302–305, 306, 307, 308, 312,
    318, 320–322, 326, 327
Free goods. See Economic free
    good
Friedman, Milton, 21

Gabyshev, K. E., 48
Gagra, 106, 156
Galanin, P. I., 84–86, 97, 99
Galazii, Gregory, 179, 183–184,
    196, 200, 202, 207
Gal'tsov, A. Ia., 92, 96
Gasoline, 132–134, 139

Geller, S. Iu., 240–242
Genossenschaften (cooperative
    water groups), 45–46
Geological work, 297–298
Georgia, 38, 72, 105–106, 137, 149,
    296
Gerasimov, I. P., 23, 110, 114, 189–
    190, 194, 196, 199, 225, 259, 262–
    263, 298
Gobi Desert, 205
Golden Sands, 155, 160
Goldman, Marshall I., 21, 107, 110,
    117, 190, 205
Golovnoi Canal, 224 (map), 248
    (map), 249–250
Gordeev, Iu. I., 258
Gorin, G. S., 84–85, 89–92
Gorky Park, 122
Gorky Street, 131
Gosplan, 39, 59, 61, 124, 146, 184,
    191
Gosstroi, 251
Gosvodkhoz, State Water Eco-
    nomic Administration, 36
Gousan, 230
Government, 3–5, 8, 21, 68, 71–72,
    161, 185–189, 246, 273–274, 281,
    290. See also Legislation; Pollu-
    tion control; State
Green belts, 299
Gresham's law of environmental
    degradation, 289
Grin, A. M., 93–94, 110, 283
Gross National Product, 70
Growth, 5–6, 29, 63–64, 66, 70, 99,
    119, 122, 188, 273, 301. See also
    Development; Industrialization
Gubakha, 138
Gulf of Finland, 100
Gulf of Kara-Bogaz-Gol, 215, 228,
    242, 248 (map), 261
Gumilev, L. N., 216
Gussak, L. A., 134

Hammer and Sickle plant, 135
Health, 37, 123, 41, 144, 148, 152,

Health (*continued*)
294, 300, 306, 307, 311, 316, 320, 326
Heat island effect, 135
Housing, 100, 104–108, 126, 139, 169, 172
Hunting. *See* Wild animals
Hydroelectric, 65, 73, 118, 157, 185, 255, 264
Hydrotechnical system, 92

Ianovskii, A. G., 93–96, 101
Ianshin, A. L., 28
Ideology, 10, 40, 44, 47–48, 51, 78, 112
Ili River, 58–59, 62, 241, 248 (map)
Ilychit, 236
Imperialism, 270
Import, 281
Incentive, 67, 117
Incineration, 174
Industrial revolution, 119
Industrialization, 5, 10, 38, 65, 82, 213, 218, 273, 279. *See also* Development; Growth
Infectious disease, 106
Infrastructure, 88, 220
Innovation, 146–147
Institute of Atomic Energy, 143
Institute of Oceanography, 159
Internalization of external costs, 45
International implications, 284–291
International public opinion, 290
Inversions, 135, 141–142
Investment, 279
Iordanskiy, A., 217, 230
Iran, 245, 248 (map), 285
Irkut River, 184, 202
Irkutsk, 149, 180, 184
Irrigation, 4–5, 16, 31, 55–56, 58, 66, 81–82, 111, 113, 118, 164–165, 221, 222 (map), 223, 224, (map), 235, 240–242, 248 (map), 249–250, 254, 259–261, 265, 268, 270,

Irrigation (*continued*)
282–283, 293, 294, 302, 303, 315, 320
Irtysh River, 114, 247, 248 (map), 250–251, 256
Irtysh-Karaganda Canal, 119, 256–257, 265
Iset River, 108
Istre River, 89, 92
Ivanchenko, A. A., 83–84, 105–107, 109, 111, 119
Ivanov, I. T., 105–106
Ivanovskaia, T. M., 142
Izmailovskii Park, 98 (map), 123–125

Japan, 6, 168, 269–270, 274, 279, 286, 290
Jet, 155. *See also* Supersonic transport
Johnston, Dr. Harry, 288

Kaliningrad, 137
Kaliuzhnyi, D. N., 110, 142
Kama River, 109, 222 (map), 248 (map), 251, 253–256
Kama River Truck Plant, 130
Kapchagaisk Dam, 58–62
Kaposhin, T. S., 105
Kapp, Karl William, 21
Kara Sea, 286
Karabash Forest Preserve, 140
Kara-Bogaz-Gol. *See* Gulf of Kara-Bogaz-Gol
Karaganda Canal, 250
Karaganda River, 137, 141, 256
Karakul' Oasis, 223
Karakum Canal, 224 (map), 235, 260
Karpov chemical and pharmaceutical plant, 126, 135
Kazakhstan, 58, 79, 114, 171, 256, 297
Kazalinsk, 248 (map), 249–250
Kazalinsk Canal, 224 (map), 249
Kazan, 94

Kazantsev, N. D., 16–17, 30–31, 34–
35, 37–39, 134–135, 161, 231,
293–299
Kes', A. S., 225, 241
Khachaturov, T., 109, 114–115
Kharkov, 106, 108, 283
Kherson, 171
Khodyrev, N. A., 157
Khosta, 158
Khram Vozdukha. See Temple of
Air
Khrushchev, N., 6–7, 30, 62, 113–
114, 129, 147, 155, 171
Kiev, 86, 93, 141, 276, 283
Kirgizia, 38, 106, 149, 223, 297
Kishnev, 105
Kislovodsk, 66, 163–166, 180
Kleimuk factory, 135
Kneese, Allen V., 53
Kolbasov, O. S., 15, 17, 28, 29, 31,
39, 40, 94–95, 105, 112, 234, 278,
283, 293–295, 297–298
Kolkhozy, 108, 167, 283, 299, 302,
304–305, 309, 320
Kolva River, 254 (map), 255
Kondratenko, A., 170
Kop'ev, Sergei, 135, 277
Koronkevich, N. I., 93–94, 110, 283
Korzhenevskii, I. B., 158, 160
Kosarev, Alexei, 217, 236
Kostylev, Valentine, 200
Kozhukhovskoi, 99
Kozlov, V. M., 144
Krasnodarsk, 156, 161
Krasnogorsk, 148
Krasnogorsk Chemical Plant, 37
Krasnouralsk, 142
Krasnovodsk, 230
Kravinko, I. V., 170
Kremlin, 98 (map), 101
Krinshch, 159
Kuban, 109
Kuibyshev, 222 (map), 230
Kum-Manych Trough, 245
Kur'ianovskii, 98 (map), 99, 101
Kuzbass mining regions, 173

Kuznetsk, 137
Kuznetsov, E., 187
Kuznetsov, I. A., 114–115, 255, 269
Kyshtym Forest Preserve, 140
Kzyl-kul, 106

Laissez-faire, 21
Lake Baikal, 7, 54, 57, 63, 66, 68,
168, 176–210, 247, 248 (map),
274
defenders of, 183–185, 191, 193–
195, 198
laws about, 31, 189–190, 192, 197,
205–210, 298, 300
natural characteristics of, 34, 54,
179–181, 195–196, 202–203, 205
proponents of economic development of, 186–188, 190–191, 193,
207
Lake Balkhash, 58–62, 241, 248
(map)
Lake Kubinskoye, 255
Lake Ladoga, 222 (map), 252
(map), 255, 269, 286
Lake Laga, 255
Lake Onega, 222 (map), 286
Lake Sevan, 65–66
Lamakin, V. V., 179
Lamb, Hubert, 263
Land, 5–7, 12, 31, 57, 71–75, 150–
175, 221, 228, 257–258, 261. See
also Agriculture; Black Sea;
Drainage; Erosion; Flooding;
Forests; Irrigation; Preserve;
Reclamation; Salination; Soil
free good, 157–160, 174, 225, 258
law, 293–294, 299, 302–303, 305,
307–308, 312, 315, 318, 324, 328
legislation, 16–17, 30, 97
salination, 259–261
virgin lands, 171
Landslide, 160, 170, 205, 303, 307,
328
Lange, Oscar, 22, 44
Latvia, 32, 296, 300

Law. *See* Legislation
Leaf burning, 137
Legislation, 4–5, 10, 16–18, 22–41, 70, 78, 94, 114, 161, 189, 229–230, 248 (map), 256, 274, 293–330
Lake Baikal, 197, 205–210
Lena Basin, 80
Lenin, 12–18, 22, 24, 28–30, 165, 167–168, 273
Leningrad, 93, 100, 104, 108, 137, 154, 222 (map), 276, 283–284
Leporskii, D. V., 282–283
Lesnikova, N. P., 38
Leukemia, 144
Likhororka River, 101
Limnological Institute, 182, 200–201
Lithuania, 38, 149, 296
Litter, 231
Liuberetskie Sewage Farm, 86, 98 (map), 99
Liublinskie Sewage Farm, 86, 98 (map), 99, 101
Livchak, Joseph, 135, 277
Log floating and rafting, 204–206, 209, 299. *See also* Forests
Loiter, M., 46, 110–111, 113–115, 118
Luddites, 267
L'vovich, A. I., 102, 106
L'vovich, M. I., 73, 78–80, 83–84, 97, 110–111, 260, 283
Lysenko, V. A., 109, 148

Magnitogorsk, 137, 142
Makeenko, M., 107
Makeyevka factory, 69
Makhachkal, 230
Malaria, 235
Malinkevich, L. B., 216
Mamaia, Rumania, 155
Mamykina, V. A., 160
Manager, 3, 35–36, 47, 67–68, 93–94, 103–104, 116, 146, 187–188, 200
Manganese, 126

Mangyshlak, 227 (map), 230, 237, 250
Marfin Brod, 91
Marginal cost. *See* Cost
Market, 3, 51, 104, 139, 213
Markizov, V. I., 91, 93, 101–103
Marx, Karl, 11–13, 18, 22, 166
Marxist ideology, 4, 40, 47–48
Maryan, Yuri, 65–66
Matskevich, V. V., 275
Mayakovsky Square, 131
Mazanova, M., 81, 115, 119
Medunin, A. E., 287
Mercury poisoning, 45
Metallurgy, 136–140
Mezentsev, V. S., 262
Micklin, Philip, 255
Military industrial complex, 289
Military power, 279
Mills, 110, 125
Minbulak, 224 (map), 249
Mineral content, 26, 61
Minerals, 302–303, 318, 327
Mining, 48–51, 166, 298, 312, 327
Ministerstvo Vneshnei Torgovli, 168
Ministry of Agriculture, 39, 275
Ministry of Defense, 159
Ministry of Electrical Energy, 35
Ministry of Geological Work, 298
Ministry of Land Reclamation and Water Management, 39, 189, 191–192, 197
Ministry of Oil Refining and Petrochemical Industries, 147
Ministry of Power, 59, 60
Ministry of Public Health, 123
Ministry of Pulp and Paper Industry. *See* Ministry of Timber, Paper, and Woodworking
Ministry of State Construction. *See* Gosstroi
Ministry of Timber, Paper, and Woodworking, 57, 186–188, 192–194, 196–199, 201–202, 206–207
Ministry of Transportation, 161

Minsk, 105, 141
Misallocation of resources, 265
Mishan, E. J., 21
Moisture, 79, 203, 302, 320
Moldavia, 80–81, 105, 296
Money, 148–149, 161, 238
Mongait, I. L., 86
Mongolia, 205, 248 (map)
Moskovorets, 28
Moscow, 28–29, 58, 60, 84–104,
    108, 122–136, 143–145, 148, 184,
    221, 222 (map), 225, 248 (map),
    252 (map), 276, 278, 280, 283,
    298
  Regional Municipal Councils, 28
  Sanitary and Epidemiological
    Station, 125–126
Moscow Oblast, 283
Moscow Oka River Basin Inspec-
    torate for Water Utilization and
    Conservation, 101
Moscow River, 85, 88, 90, 92, 96–
    97, 98 (map), 101, 102, 222
    (map)
Moscow–Volga Canal, 90, 92
Moskva River. See Moscow River
Mosquito, 235
Mozhalsk, 92
Muinak, 236
Muscovites, 85
Muskrats, 56, 246
Mytushchinskii, 84

Nace, Dr. Raymond L., 263
Nagibina, T., 35
Narzan, 166
Nationalization, 4, 8, 75. See also
    Government
Natural gas, 150, 281–282, 327
  recovery, 48, 127–128, 130, 156,
    281–282
Natural resources, 294
Nazarov, I. A., 282–283
Neighborhood effect. See Exter-
    nality.
Net National Well Being

Net National Well Being (cont.)
    (NNWB), 70
Neva, 100, 104
Night soil, 85
Ninth Five-Year Plan, 169, 255–
    256, 279
Nivel Izhem, 253
Nizhnii Tagil, 137
Non-Marxist, 4, 18, 22
Norm, 110, 135, 165, 201–202, 208,
    303, 320–321, 326
North Dvina, 109
Novogork'y Oil Refining, 37, 278
Novokuznetsk, 138
Novoselov, E. N., 179, 196
Nuttonson, M. Y., 37, 123, 125,
    129, 135, 137, 139–140, 141–142,
    145
Nuclear weapons. See Atomic
    testing; Atomic Test Ban Treaty

Ob River, 247, 248 (map), 251,
    256, 259, 263
Oceanographical Commission of
    Academy of Sciences, 241
Odessa, 86, 106, 108–109, 138, 155,
    160, 283
Oil, 48–52, 71, 102, 110, 115, 127,
    139, 147, 163, 189, 207, 229, 230,
    231, 236, 243, 257, 258, 282, 284,
    286–287, 326
Oka River, 92, 101, 109, 222 (map)
Oldak, P., 53–54, 70
Onega River, 222 (map), 255
Opportunity cost, 163, 172, 259
Ore, 48, 141
Orenburg, 68, 106
Ostankino, 94
Outhouse, 105
Ovsiannikov, N. G., 101, 189
Ozern Reservoir, 91
Oziranskii, S., 80–82

Pacific Ocean, 246, 269–270
Pakhrinskii, 98 (map), 100
Panov, D. G., 160

Paper and pulp industry, 20, 56, 57, 168, 181, 183–184, 186–190, 192–193, 202, 204, 285. *See also* Baikalsk Cellulose Carton factory; Selenga Cellulose Paper factory
Parks. *See* Public park
Party, 23–24, 208
Pavelets railroad station, 94
Pavlosk, 137
Payback period, 57, 220, 255, 266
Peasant, 113–115, 168–169, 171, 260
Pechora-Kolva Canal, 253
Pechora River, 222 (map), 247, 248 (map), 251, 253, 254 (map), 255, 263
Perekop River, 270
Petrianov, I., 46, 137, 140, 145, 147–150
Petroleum, 142, 327. *See also* Oil
Petrovsky, B. V., 23–24
Phosphorous, 102
Piatagorsk, 164
Pigou, A. C., 19–22, 44. *See also* Externality
Pinsk, 105
Pitsunda, 157, 159
Plan, 67, 69, 95, 138, 147, 164, 171, 181, 183–184, 194–195, 197, 199, 212, 240, 251, 301, 309, 311–313, 320, 326, 329. *See also* Five-year plans; Gosplan; Target; Priority
Planners, 6–7, 127, 145–146, 155, 163, 183, 186, 192, 195, 213–214, 220, 244, 246–247, 251, 256–257, 260, 266, 269, 279, 282
Podkumok River, 164, 166
Pokcha, 253, 254 (map)
Polevskoy, 140–141
Polevskoy cryolite plant, 141
Pollution control, 3, 6, 22, 24, 26, 35, 67–70, 97, 116, 126, 130, 134–135, 139–140, 146–150, 187–188, 242, 272, 275, 290. *See also*

Pollution control (*continued*) Legislation
Popov, N., 23
Popov, V. I., 107
Population, 5, 11, 23, 79, 81, 88–89, 91, 100–101, 105, 107, 142, 154, 233, 251, 302–305, 308–309, 311–312, 315–316, 318–320, 323, 326, 329
Potassium recovery, 48
Poti, 156
Powell, David, 37
Precipitation, 78–79, 218
Premiums, 35, 65, 68, 165, 299 denied without sanitary treatment, 39–40
Pre-revolutionary period, 84–87, 97, 247, 251, 273, 282
Present discounted value, 49
Preservation of nature, 296–297. *See also* Law
Preserves, 273–275, 294
Price, 21–22, 51–52, 57, 71, 112, 127, 132, 162, 169, 171–172, 225, 258, 260, 279–280 as scarcity relationship, 55, 73, 115
speculative, 233
Price reform of July 1967, 51, 55, 258
Priority, 68, 136, 233, 251, 280, 315
Private enterprises, 146, 186, 188
Private greed, 3, 4, 186, 208
Private land ownership, 70–75, 273
Private profit and pollution, 3–4
Private property, 70–75, 186, 273 as check on pollution, 162–163, 214
Production ethic, 66
Profit, 56, 67, 146, 187–188, 212, 242
maximization, 48, 187
Proshchenko, V. F., 171
Pshad River, 159
Psychic cost, 266

Public accountability, 188–189
Public health. *See* Health; Sanitation
Public interest, 2, 3
Public ownership. *See* State
Public park, 273–274, 293, 305
Public relations, 27
Public works, 268–269

Radioactivity, 128, 142–144, 192, 287–289
Radiology, 237
Railroad, 156, 159
Ramenskii, L. A., 138
Raw materials, 13, 50–52, 72, 118, 150–175, 257–258, 261, 280–281, 304–305
 as free good, 258
 for foreign exchange, 64
Reclamation, 173, 189, 191–192, 197, 268, 306–307, 320
Recovery rates, 48, 50
Recreation, 56, 58–59, 60, 154, 266, 311
Recycling, 13, 82, 94–96, 110–111, 139, 145, 174, 184, 198–199, 279–280, 282–283, 321
Regulation. *See* Legislation
Rent, 12, 50–52, 169
Reservoir, 91, 119, 221, 222 (map), 224 (map), 247, 248 (map), 249, 253, 258–259, 261–262, 268, 298, 312, 315, 321–322, 324, 327. *See also* Dam
Resorts, 29–30, 73, 97, 106, 139, 154–158, 160, 163, 295, 300, 302, 306–307, 316, 320
Revolution of 1917, 24, 66, 88, 100, 119, 168, 311
Riad Grazdanskaia Prospect, 100–101
Riazan oil refining, 115
Ribas, Admiral, 251
Riga, 180
Risk, 69
Rivers, 20–21, 34, 37, 45–46, 55,

Rivers (*continued*)
 58, 62, 69–70, 78–80, 83, 85, 89–93, 96–97, 100–102, 104, 106–109, 114–115, 156–161, 164, 179, 182, 184, 203–204, 212, 218, 220–221, 222 (map), 223, 224 (map), 230, 233, 240–241, 247–251, 252 (map), 253, 254 (map), 255–257, 267, 270, 302, 312, 321–323, 327
 dead, 232
 psychology of, 220
 reversing flow of, 246–258, 261–262, 264–266, 286
Romanenko, A. M., 115
Rossolimo, L., 179–180
Rotating earth, 263–264
RSFSR, 32–33, 81, 108, 169, 189, 297
Rubleva water works, 85
Rublevsk, 88–89, 91
Ruhr, 45–46, 110, 116
Rumania, 155, 161, 285
Runoff, 78–79
Ruz Reservoir, 91
Ruz River, 91, 92
Rybnits, 106

Salination, 225, 234–235, 245–246, 259–261, 303, 315, 328
Samoilov, Ia. M., 107
Samuelson, Paul, 21
Sanitary brigade, 85–86
Sanitary inspector, 192
Sanitary protection zones, 294–295
Sanitation, 148, 190–192, 293–294, 298, 300, 303, 316, 319–320
Sarkisian, S., 183
Sarukhanov, G. L., 222, 255
Satke, 140
Seismic explosion, 229
Selenga Cellulose Paper factory, 189–190, 193–194, 196, 202, 206–207, 209
Selenga River, 182, 184
Sennikov, V. A., 179–180

Serova, O., 183
Sever water plant, 190–191
Sewage, 13, 17, 30–31, 37, 45, 82–
    129, 124, 182, 184, 191, 200–201,
    208–209, 230–231, 282–283, 285,
    294, 298, 304, 316–318, 320–321,
    324–326
Shabad, Theodore, 29, 256–257,
    295
Shchekino Plant, 61, 138
Shchelkovskie line, 98 (map)
Shipping, 322
Shiskin, N. I., 251
Shitunov, F. Ia., 137, 203
Shkatov, V., 116–117, 136
Shnitnikov, A. V., 215–216
Shul'ts, V. L., 54–55, 216–217, 223,
    238, 240–242
Siberia, 79–80, 169, 178, 183, 205,
    240, 247–251, 253, 255–259, 261–
    262, 266–267, 286
Silt, 168, 203, 234, 243–244, 327
Simonov, Anatoli, 229
Sinel'nikov, V. Ye., 88, 96, 102
Sitnin, V. K., 52
Sixth Scientific and Technical
    Conference on Sewage Treat-
    ment, 100
Slag, 50
Slelekhov, 149
Sludge, 102, 109, 283
Smith, Stephen C., 53
Smog, 125, 134
Smoke, 122, 125–126, 135–136, 138–
    139, 150, 275, 282
Sochi, 155–158
Social cost. See Externality
"Socialist competition," 165
Society for the Protection of
    Nature, 189
Soil. See Land; Erosion
Sokol'niki Park, 17
Sokolovskii, M. S., 25, 122–124,
    126–128, 129, 130, 131, 132
Soloukhin, Vladimir, 68–69
Solzan River, 184
Sovkhozy, 108, 167, 302, 304–305,

Sovkhozy (continued)
    309, 320
Sovnarkhoz, 139
Speculators, 71
Spillover. See Externality
SST. See Supersonic transport
St. Petersburg, 84, 86, 252 (map)
Stalin, 28–29, 89, 155, 167, 171,
    220–221, 225
Stas', I. I., 228–245
State, 22, 69, 89, 94, 165, 273
    advantages of, 270–291
    budget, 113
    as de-emphasizer of consumer
        goods, 279–282
    ownership 311–313
    as producer and chief polluter,
        2–4, 7–8, 10, 20, 33–34, 44, 46,
        58, 69–70, 73, 181, 208, 214, 275
    as zoner, 278–279
State Committee on Prices, 51–52
State farm. See Sovkhozy
State Institutes for the Design of
    Cellulose and Paper Plants
    (Gipprosbum), 184
State Sanitary Inspectorate of the
    Ministry of Health, 39, 190, 299
Steam, 126, 144, 276, 277, 282
Steam and Electric Utilities, 144
Steam-generating plant, 68, 95,
    136
Stechken, Boris, 132
Steel, 70, 110–111, 137
Stravropol, 94
Streams. See Rivers
Streeten, Paul, 57
Strip mining, 172–173
Subsidy, 144, 264
Sudak, 160
Sukhona River, 222 (map), 255
Sukhotin, Iu., 48, 115, 258
Sukhumi, 155–156
Sulfur, 127–128, 133, 139
Sulfur anhydride, 141
Sulfur dioxide, 125, 139, 141
Sulfur oxide, 123, 128–129
Superphosphate, 148

Supersonic transport, 7, 287–289, 290
Supreme Soviet, 23
Sushkov, T. S., 27, 32–33, 81, 147–148, 150, 169, 297
Sverdlosk, 106, 108, 137, 142
Synthetics, 5, 63, 110
Syr Darya River, 55, 62, 218, 223, 224 (map), 241, 248 (map), 249–250, 267

Tadzhikstan, 223, 296
Tallin, 131
Tambukan, Lake, 166–167
Target, 7, 35, 65, 68–70, 127, 147. See also Plan
Tashkent, 93, 224 (map), 248 (map)
Taxes, 113–115
Taylor, Fred M., 22, 44
Tbilisi, 137
Technology, 7, 11, 83, 178, 214, 220, 247, 267, 288–290, 308, 321
Temperature. See Thermal pollution
Temple of Air, 164
Teplovaia Elektrotsentral' plants (TETs), 75, 126, 135–136, 141, 144–145, 275–277
Thermal cooling, 80
Thermal pollution, 26, 81, 95, 216, 237, 262–263, 270, 276–277, 303
Thermal use, 81
Thermopalae, 164
Tikhomirov, F. K., 255, 269
Timber. See Forests
Time horizon, 256
Timoshchuk, V. I., 236–237
Tiraspol, 106
Tiumen, 106
Tobol River, 247, 248 (map), 249, 251
Tobolsk, 247, 248 (map), 249–250, 256
Tolstoi, M. P., 283
Tolstoy, Leo, 61–62, 138

Tourist, 154, 165
Trofiuk, A. A., 31, 110, 189–190, 184, 196, 199, 298
Troshkina, E. S., 205
Tsuru, Shigeto, 70, 268
Tsuru's complaint, 268
TU 144. See Supersonic transport
Tuapse, 156, 158
Turgai Gates, 248 (map), 249
Turkestan, 293
Turkestan Canal, 224 (map) 249–250
Turkey, 248 (map), 285
Turkmenia Canal, 224 (map), 250
Turkmenistan, 37, 250, 297
Turnover tax, 279
Tyuyamyyan, 223

Uchinsk Reservoir, 90
Ude River, 203
Ukhta, 257
Ukraine, 79, 94, 109, 150, 169, 171, 282, 296
Ukranian Ministry of Ferrous Metallurgy, 69
Ulan Ude, 182, 209
United States, 2, 3, 5, 7, 8, 34, 38–39, 44, 45, 47, 53, 73, 78, 82, 84, 89, 104, 107–110, 117–119, 127, 129–131, 134, 137, 140, 146, 147–150, 152, 154, 162, 170, 172–174, 179, 186, 188, 194–195, 212–213, 237, 263–266, 269, 273–274, 276–278, 279, 282–284, 286–288, 290–291
Ural River, 109, 222 (map), 230, 232–234, 243, 248 (map)
Urals, 140, 195, 221, 248 (map)
Urbanization, 5, 6, 7, 20, 157, 169, 218
Urea plant, 62–63
Ust Kulom, 253, 254 (map)
Ust-Urt Canal, 224 (map), 250
Uvod' River, 106
Uzbekistan, 38, 149, 223, 296

Vain, A., 52
Vainshtein, A., 47

Value, 162–163, 171–174, 202, 225, 258, 281
Varna, 155
Vasil'eva, E. M., 13
Vazuza Reservoir, 92
Vendrov, S. L., 158, 160, 228, 261, 266
Ventilation, 13, 137
Vested interest, 3, 74
of local officials, 36
of private property owner, 70–75
Violation of law, 192–193, 197
Vitt, M., 35, 278
Vladimir, 106
Vol'fson, Z. V., 131
Volga Canal, 93
Volga-Don Canal, 222 (map), 225
Volga River, 34, 37, 70, 90, 92, 106, 108–109, 218, 221, 222 (map), 225, 227 (map), 230–236, 242, 245–246, 248 (map), 250–251, 253, 256, 264, 267, 275, 285
on fire, 231
Volkov, Oleg, 185
Volodicheva, N. A., 205
Vorenezh, 106
Voskresensk chemical plant, 37, 278
Vostok water plant, 90
Vozha Lake, 255
Vyborg district, 100
Vychegda River, 222 (map), 247, 251, 253, 254 (map), 255

War communism, 86
Wastes, 6, 13, 20, 23, 44, 83, 94, 105, 139, 143, 145, 149, 173–175, 182, 183, 185, 192, 200, 230, 231, 279, 280, 282, 283, 298, 308, 326
Water, 5–7, 14, 23, 28, 35–36, 45–46, 54, 58–59, 61, 65–66, 69–70, 76–120, 122, 168, 173, 202, 212, 220, 242, 278, 284–285, 287. See also Aral Sea; Caspian Sea; Dam; Evaporation; Irrigation; Lake Baikal; Rivers; Reservoir; Sewage
circulation of, 199

Water (continued)
free good, 47–48, 110
laws, 294–295, 298, 300, 302–304, 307–308, 311–330
legislation for, 17, 24, 26–27, 30–32, 39–40, 114, 147–148, 205
owned by state, 311–312
price of, 260–261
psychology of, 214–215, 220–221, 266
supply. See Reservoir
surplus, 246–247, 258
underground, 31, 261, 297–298, 302, 307–308, 312, 320, 327
wells, 294
Weinberger, Leon W., 82–84
White Sea, 221, 222 (map) 269
Wild animals, 17, 30, 294, 297, 305, 307
Wiles, Peter, 214–215
Wind circulation, 237
World War II, 3, 28, 99, 113, 122, 138, 164, 218, 223, 271

Yakolev, M. P., 101–102
Yaksha, 253, 254 (map)
Yalta, 51–52, 155, 156, 157, 160
Yasnaya Polyana, 61–62, 138
Yausa River, 92, 98 (map), 101
Yenesei River, 247, 248 (map), 251, 263
Yermak, 256

Zakrestov, 99
Zapad water plant, 91
Zemstvo. See Moscow Regional Municipal Council
Zenkovich, V. P., 156, 158, 159–160
Zhdanov, A. M., 156, 158, 160, 162, 283
Zheleznoi Mountains, 166
Zhukhov, A. I., 86, 119, 171
Zhuvkovich, L. A., 265
Zile, Zigurds L., 15–18, 30, 70, 293–294
Zinger, N. M., 276
Zolataya Gora, 140
Zykova, A. S., 128

TD
187.5
R9
G63

Goldman, Marshall I.

The spoils of
    progress:
    environmental
    pollution in the
    Soviet Union

DATE		
MAR 2 1981		
DEPT. LOAN		